LIFE & SPACE

LIFE & SPACE

Philosophical and scientific fragments

Caden James Howlett

First edition, 2024

ISBN: 9798394266355

For my parents

You can't control your thoughts, you can't control your feelings. Because there is no controller. You are your thoughts and your feelings. And they're just running along, running along, running along. Just sit and watch them. You're still breathing aren't you?

--Alan Watts.

Acknowledgements

Writing a book is a challenging and time-consuming process that requires spending a lot of time alone in a quiet room. However, even more important than the hours put in is inspiration to start in the first place. I am fortunate to have been surrounded by many intelligent, open-minded, and loving people during the writing of this book. Among them, I owe the utmost gratitude to Zach Schmidt, who has supported me with unwavering friendship and encouragement during and on either side of the process. Countless hours of conversation with Josh Plovanic and Chance Ronemus helped me refine the ideas presented in this book.

Other individuals who helped and inspired me during writing include Aislin Reynolds, Alexis Ault, Andrew Mullen, Andrew Pollard, Barbara Carrapa, Ben Hebert, Christopher Kussmal, Cody Grant, Colin Rossano, my sister Danielle Howlett, Devon Orme, Drew Laskowski, Dylan Branscum, Emilia Caylor, Emit Meyer, Erica Duncan, Eytan Bos Orent, Gilby Jepson, Haiyang Kehoe, Haley Thoresen, Holly Thomas, Ian Dodds, Jacob Gardner, Jared Sillanpaa, Jenna Biegel, Jon McKane, Ken Gourley, Kurt Sundell, Lauren Reeher, Matthew Yaeger, Mike Bournstein, Misia Zilinski, Owen Silitch, Pete DeCelles, Phia Swart, Priscilla Martinez, Robert Hayes, Sophia Bautista, and Tshering Lama Sherpa.

First, last, always, and in a number of words inversely proportional to my gratitude, I thank my parents, Julia and Howie.

Contents

Introduction

When I fell in love with astronomy, the feeling was unmatched. I remember gazing through the winter night skies in Montana realizing how little I knew and how badly I wanted to know. Crystal clear atmosphere, biting cold. Milky Way. Equipped with no knowledge of space, my mind didn't have anywhere to go, and all I could do was look. It was a strange feeling of joyful helplessness. All who have contemplated the immensity of space have felt this at some point.

The progression from such wonder to education is natural. Questions like, "How do stars create light?" and "Where is the edge of space?" arise organically. The only way to answer such questions is through study. Fortunately for us, millennia of rigorous scientific research—confronting topics ranging from wormhole formation to the geology of Mars—is condensed into books and available at our fingertips. And so those filled with wonder take a dive. Propelled by curiosity, we swim through books, music, art, podcasts, and film in an attempt to understand. Perhaps that is why you hold this book now...

The more I learned about the cosmos, the more I realized I didn't understand. As a result, many of my questions funneled back to our planetary neighbors and planet Earth itself. Questions like, "Why are the gas giant planets farther from the Sun?", and "How do mountains form on Earth?" infiltrated my cranium. The funnel narrowed when I realized how heavily we rely on our limited human senses to interpret the cosmos. I was bothered that the thing that makes scientific observation possible—consciousness—was completely blanketed in mystery. Naturally, the questions, "Is consciousness a physical process?" and, "Do our senses actually provide a window into objective reality?" came barreling in.

An interest in astronomy took me from the edge of outer space to the space between my ears. For me, it underscores a point simple and emphatic: we are surrounded at all scales of observation by fundamental and deeply important unanswered questions. Many of these can be probed using the scientific method, many are currently out of reach, and some may be truly impossible to answer. Regardless, it is my opinion that there is no better (or more fortunate) way to spend a life than by attempting to understand the complexities of existence. I've just begun this endeavor.

This book contains a selection of written thoughts I've had during my journey from joyful helplessness under the night sky to today (still joyful and helpless). Broken into five parts, I will gradually increase the scale of our contemplations from the human to the infinite. Part I (*Human*) considers Stoic philosophy, neuroscience, and mortality (among other things). More broadly, we will think about the human experience and how it relates to our place in space. Part II (*Civilization*) is the shortest part and includes (but is not limited to) essays on energy consumption, science and religion in society, and collective human experience. Part III (*Earth*) contains ponderings on topics in geoscience (both rock and water) and biology and conservation. Part IV (*Cosmos)* takes us into the night sky where we will focus on the solar system, black holes, time warps, stellar evolution, and more. Part V (*Beyond)* ventures into the realms of futurism, artificial intelligence and the simulation argument, consciousness, and metaphysics. The borders separating each part are blurred, but I have done my best to arrange the essays in an order that makes the book more than the sum of its parts. I have kept the delivery casual, kinda like a journal—consider it a sign of friendship.

Life is too short for long introductions. Let us begin our journey through mind, material, space, and time.

· · · · ·

Part I: Human

The default mode network

When you think too much about something, it carves a groove in your mind. Whether it's about a person, a past occurrence or an event in the future, overthinking will incise a channel in your brain that can be difficult to escape. Although this may sound like a simple metaphor spoken by a hopeless romantic (perhaps partially true), there is a neurological basis for these claims.

In neuroscience, there is a set of interacting brain structures known as the *default mode network* (DMN). The DMN links parts of the cerebral cortex (the most recently evolved part of our brain) with deeper and evolutionarily older structures that partially control our memory and emotion. Serving as well-established neurological highways, the DMN is most active when the mind is wandering, which is usually the time when we are thinking about others, thinking about ourselves, or thinking about the past or future (i.e., time travel in the brain). These linkages have been documented using neuroimaging techniques like functional magnetic resonance imaging (fMRI), where brains in a passive state have heightened DMN pathway activity. Similar to how a river channel becomes more established as water is discharged through it, the DMN is strengthened by the activation and passage of thoughts. Put another way, a fixation on a certain thought strengthens the specific pathways that facilitate their passage.

The neuroscientist Dr. Mendel Kaelen proposes a beautiful and relevant metaphor which involves picturing the brain as a snow-covered hill, with thoughts being sleds gliding down it. As more and more sleds glide down the hill, they are inevitably drawn into preexisting tracks. This magnet-like effect represents our thoughts being pulled into the most well-traveled neural connections in our brains, and over time, it becomes more difficult to take any path other than the ones already established. This is how overthinking something becomes a positive feedback loop where the more you think about it, the more difficult it is to escape.

We've all been caught in a track before, and maybe some of us are currently stuck in one, but continued neuroscience research is helping us identify ways in which we can temporarily flatten the snow and create a mind that is more impressionable and free-thinking. The figure shown here is extracted from Michael Pollen's book *How to Change Your Mind*. It illustrates the communication between brain networks in placebo recipients (a) and people given psilocybin, the naturally occurring psychedelic compound in so-called "magic mushrooms" (b).

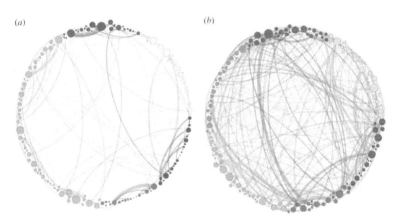

Communication between brain networks in people given psilocybin (right) and a non-psychedelic compound (left). From Petri et al. (2014).

When it's dark out

When you fall asleep tonight, blood will flow out of your brain and your neurons will settle down. Shortly after, a watery liquid will infiltrate the nooks and crannies of your brain in rhythmic, pulsing waves. This liquid, known as cerebrospinal fluid, flows through the brain during sleep and washes toxins and other metabolic garbage out. In other words, your brain becomes a dishwasher (eight-hour cycle, no heated dry!!). Matthew Walker's fantastic book titled *Why We Sleep* gave me a new appreciation for our nightly slumber. His detailed overview of modern sleep-related neurological studies makes clear that good sleep is (almost without question) the most important component to living a long, healthy, and fulfilling life. The phenomenon of "brainwashing" outlined above is one of the many incredible things that happens to us when the lights go out. I highly recommend the book.

A bottle of wine brought my friends and me to the question of what happens to us when we fall asleep. We had been discussing the nature of consciousness, so the question was approached from this angle (i.e., where does our consciousness go when we are asleep?). Similarly, does our consciousness continue while we sleep, or does it stop and then restart? Considering these questions naturally brings up another interesting one: is the person who wakes up in the morning the same person who went to sleep the night before? This is a totally valid philosophical question, and some have argued that it would be impossible to test whether our morning selves are actually us, or a different person with identical memories. There is no strong evidence to suggest that there is some mental "thing" that exists continuously during our lives, and since our conscious experience is broken during slumber, an argument can be made that tomorrow morning we will not be the same person. An unsettling thought with interesting implications for how we perceive our lives and the illusory self that seems to be riding around inside of us.

Levels of conscious experience

Homo sapiens consider themselves the most important species on Earth because they have become conscious. When contemplating consciousness, it is interesting to consider how and why we define it the way that we do. We loosely define consciousness as a state of being aware of one's surroundings. The word "aware" that is tucked within creates problems since there is a lot of room for further interpretation. What does it mean for a species to be aware? Why should we be the species fortunate enough to set the bar of awareness? We define consciousness based on our subjective experience, and just because we know what it feels like to be a human being doesn't mean that we have reached true awareness.

There is no reason to think that our species lies at the pinnacle of evolution. Therefore, there is no reason to think that we have reached the pinnacle of conscious experience. On the contrary, it seems likely to me that we are a long way off from true awareness. If you think about the evolution of consciousness over time, it is fair to say that we currently sit close to its origin. If conscious awareness continues to heighten over time, we can conclude that we are far from its limit (if there is a limit). This has interesting implications for our current understanding of what it means to be conscious. For example, if we create a superintelligent artificial mind, it could plausibly scale the ladder of awareness to far beyond what we currently understand. *Homo sapiens* left in the dust. On its perch on higher awareness, the artificial intelligence may look down on our species as we currently look down on mosquitoes or algae.

Perception ≠ reality

Alongside the nature of consciousness, limitation of human senses is one of my favorite topics of consideration. As I explore these topics, I realize that they are interconnected and must be discussed together.

In my leisurely reading on the senses, I came across the thought-provoking work of Donald Hoffman. Dr. Hoffman is a cognitive psychologist and computer scientist who is known most prominently for his work developing a theory of human perception. In essence, he is interested in figuring out how and why we have conscious experience, and what role consciousness plays in our perception of the world around us. Exactly what

I've been looking for! The hypothesis that he is actively testing and in support of can be explained no better than by he himself: "...perceptual experiences do not match or approximate properties of the objective world, but instead provide a simplified, species-specific, user interface to that world."

The huge statement above is rooted in what is known as *evolutionary game theory* (EGT). To understand EGT, one must first understand game theory more broadly. In game theory, we consider how and why people (or animals, computers, etc.) make the decisions that they do in the context of known payouts or quantifiable consequences. For a simple example, we can look to the "prisoner's dilemma", which involves two hypothetical prisoners who are being questioned separately for a crime. The game can be set up as follows:

1. If both confess, they will each receive a five-year prison sentence.

2. If Prisoner 1 confesses, but Prisoner 2 does not, Prisoner 1 will get one year and Prisoner 2 will get eight years.

3. If Prisoner 2 confesses, but Prisoner 1 does not, Prisoner 1 will get eight years, and Prisoner 2 will get one year.

4. If neither confesses, each will serve two years in prison.

The rational strategy to take here would be number 4. However, the fact that neither prisoner knows what action the other will take may lead to an outcome that is statistically unfavorable. Therefore, broadly speaking, game theory tells us that the gains from cooperation can be greater than the rewards from pursuing self-interest.

EGT extends these concepts to fitness payoffs (things that you want to survive; think of them as points in a game). In the context of natural selection, the ability to gain such "points" depends on both the state of the world and the organism pursuing them. Hoffman runs EGT simulations in set environments with organisms that are programmed to "see" varying levels of what "objective reality" is (i.e., whatever objective reality is specified to be in the simulation). His results show that organisms who see objective reality consistently go extinct, offering additional support for his *interface theory of perception*.

At its most fundamental level, his theory of perception is an anti-physicalist one that states that consciousness creates physical objects and their properties. Everything in the cosmos, spacetime included, is a manifestation of consciousness. Hoffman's theory claims that the cosmos is analogous to the interface of a laptop, simply existing to help the user perform useful tasks without showing him or her the "truth" of the computer. Just as a computer interface hides the truth of the computer (circuits, software layers, etc.), could it be possible that our senses have evolved not to show us objective reality, but rather an interface to maximize our chances of reproduction? Are we radically deluded in what we consider true reality? Could it be possible that the way in which we experience fundamental matter is subjective and varies from species to species? The answer to these questions could very well be yes. I have a growing feeling that reality is far stranger than we can comprehend.

Interface theory of perception: implications for science

So, our perceptions may not be a window on objective reality, and matter and spacetime as we know them may be simple manifestations of our consciousness. Worrying ideas, but important to consider!

I have exchanged emails with Don Hoffman, as I was curious to hear his opinion on the scientific method, with special emphasis on the physical sciences. As a geologist whose entire discipline is built on the assumption that physical reality (both spatial and temporal) is close to objective reality, I was kind of disturbed by Hoffman's claim that our understanding of matter, space, and time do not even approximate reality. Physical sciences investigate physical phenomena, so doesn't taking an anti-physicalist view on the world therefore strip these disciplines of their meaning and value? When I proposed this question to Don, he replied that, "...studying the physical sciences is essential...only by such study can we get the boundary conditions on a deeper theory of reality." I agree that we currently have no better method with which to explore and eventually uncover reality, but this conclusion does still not justify the studies of many physical scientists (myself included). For example, isn't investigating the timing of crustal deformation on Earth futile if time and matter as we know them are manifestations of consciousness? Doesn't it make my studies the equivalent of saying that anything equals anything? That our interpretations are grounded in data that is delusory, data that simply arise from the sensory perceptions that evolution has given us with the sole purpose of maximizing our chance of reproduction. I suppose that all scientific work contributes in some way to pushing the limits of our species, but this theory of perception has me questioning the foundation upon which our interpretations (and science) are built.

Conscious realism

"It is a natural and near-universal assumption that the world has the properties and causal structures that we perceive it to have; to paraphrase Einstein's famous remark, we naturally assume that the moon is there whether anyone looks or not. Both theoretical and empirical considerations, however, increasingly indicate that this is not correct." –Fields et al. (2018).

Donald Hoffman's *User-Interface Theory* suggests that our perceptions do not match reality and that what all humans perceive as reality ("consensus reality") is illusory. Spacetime itself, he argues, is the manifestation of a species-specific interface that has evolved solely to increase our chances of reproducing in the future. If this is true, the very language with which we describe the physical universe is wrong, and we need an entirely new theory of reality (because objective truth must exist somewhere, right?).

Conscious realism is a new theory of reality proposed by Hoffman. It builds upon the user-interface theory and argues that (1) consensus reality is an illusion, and (2) reality is made of a complex, dimensionless, and timeless network of what he calls *conscious agents*. Conscious agents can be thought of as points of view. Hoffman references the famous experiment in which a cutting of the bridge between the hemispheres of the brain results in a single person having two separate consciousnesses inside of them (more on this later). This perhaps suggesting that we (our consciousness and attendant perception) are the result of stacked conscious agents that can be split ad infinitum (i.e., it's conscious agents all the way down). Regardless of the structure of reality, a mathematical proof is offered by Hoffman and Prakash (2014) that supports the interpretation that perceptions do not even approximate what is objectively real.

Experience is not an illusion

In a cosmos where our perceptions may not match reality and our science may not be converging on absolute truth, why do we keep going? If the world around us is an illusion, why take anything seriously? I think there is one thing that keeps me and many others from descending into a pit of nihilism.

You cannot take my experience away from me. That is, you cannot strip me of what it is like to be having conscious experience right now. Many ancient philosophers and modern scientists have made the argument that consciousness is the only thing in the cosmos that cannot be an illusion. Indeed, it is like something to be me right now, and for you to be you. Although we may be at the low end of a potential spectrum of consciousness, and people may have different definitions for what it actually is, I don't think I've ever heard a sound argument for an actual absence of it.

I had an enlightening conversation with a homie in which we concluded that even if our perceptions do not match objective reality, it does not justify taking a meaningless view on the cosmos. For the sake of conversation, let's say that we are making scientific interpretations that are grounded in data that is meaningless, and that we are delusional in thinking that they are leading us towards absolute truth. Even if that's the case, we all still know what it feels like to prove or disprove our own hypotheses. It feels good either way, and that feeling cannot be taken from us. We know that there is no end goal in life—that "life is a journey and not a destination", as Ralph Emerson once said. So why should there be an end goal in our exploration of the cosmos? It could be that the feelings we derive from science are its meaning and the only real thing about it.

Science aside, you cannot take away the love, desire, fear, and excitement that I will feel today. And those are the things are worth living for.

A strange creature

Octopus, chameleon, mushroom, giraffe...we share the planet with organisms that look more alien than any image that I can conjure up in my mind. The magnificence of our naturally selected friends, paired with how often we see other humans, causes us to overlook our own weirdness.

Repetition breeds indifference. The more we see or experience something, the less interesting it becomes. It makes sense evolutionarily to not be in constant amazement of our surroundings, but it absolutely strips away some of the magic of existence. A textbook example lies in our indifference towards other members of our species. Many of us see hundreds to thousands of human people every day of our lives—the ultimate repetition. As a result, we are usually not fascinated by the sight of another human. On the other hand, we do not often get to see giraffes or chameleons, so we are enthralled by the sight of them. I'm making the (safe) argument that humans are by definition as strange and beautiful as any other creature.

One doesn't have to think hard to recognize how fu**ing strange the human being is. Our soft and vulnerable endoskeleton is all that separates our vital organs, blood, skeleton, and brains from the external world. These bodily components are all carried along bipedally, with additional appendages waving around up above to keep balance. We take up a fair amount of space when standing, but are still mostly appendages, with a small torso hosting our organs. Our overly complex human brains add the additional baggage of anxiety, hopes, fears, loves, and desires to the mix. Of course, I could go on indefinitely describing how weird we are.

I write all this in the hope that you will approach your fellow human beings differently today. At least once, it would be cool to break through the veil of indifference and see someone (or yourself) as what they (or you) are—a very strange creature. I will do the same. In fact, even thinking about you reading this is creeping me out.

Metabolism

"Metabolism is the set of life-sustaining chemical reactions in organisms." I recently realized even though I knew this general definition of metabolism, I didn't really understand how it works. It seems ridiculous to not have at least a fundamental understanding of the processes that literally make everything else possible! All this is to say that I've been reading into the biological literature a bit and trying to make sense of these kinds of processes—it's been fun so far.

As the process responsible for maintaining life, I wanted to learn more about where metabolic energy comes from. At a fundamental biochemical level, metabolic energy is created within cells called respiratory complexes. The molecule adenosine triphosphate (ATP) plays a central role in the creation of metabolic energy, as it breaks down from triphosphate (three phosphates) into diphosphate (two phosphates). When this third phosphate bond is broken, its binding energy is released as the source of your metabolic energy, keeping you alive. This process works in reverse as we consume food and breath oxygen, converting the broken ADP back into ATP. This flux of ATP is often referred to as the currency of metabolic energy.

At any given time, our bodies contain about 250 grams of ATP. However, through the constant break down and recreation of ATP throughout the day, the human body typically makes 2.0^{26} ATP molecules (~200 trillion trillion molecules). This corresponds to a mass of approximately 80 kilograms (175 lbs). Reread dat! Each day, we produce and recycle our entire body weight worth of ATP. I find this truly remarkable.

I've always been mildly unsettled by the fine-tuned and fragile processes that keep us running; it seems like there are too many places where things can be derailed. But remarkably, we keep going through processes like that above. I know many of us spend a lot of time looking up, but I think it is equally important to look in.

Human energy consumption

The human body is extraordinarily efficient in its use of energy, with an average human requiring ~2,000 calories a day to stay alive. To put this in perspective, 2,000 calories is equivalent to just shy of 100 watts, which is approximately equal to a standard lightbulb. Think of all the natural activities we can do, both physically and mentally, with the same amount power as a household lightbulb!! It makes lightbulbs look pretty damn lame, bringing me to my next contrasting point.

Man-made objects are profoundly inefficient. Your dishwasher is a good example, requiring over ten times more energy per second than you simply to wash dishes. Your automobile is another one, which to move uses more than a thousand times more energy per second than your body. The inefficiency of our inventions has important implications relating to the energy budget and, ultimately, the environment. In the United States, each citizen (on average) uses over 100 times more energy than their natural requirements each day (somewhere around 11,000 watts). This amount of power is about equal to the metabolic rate of a full-grown blue whale, which has a mass >1,000 times ours. A fact worth repeating: each individual in the US is consuming the same amount of energy as a creature 1,000 times our size.

A final implication of these remarkable facts is that the human population is operating as if it were much larger than its eight-billion people. With a global average for individual energy consumption hovering around 30 times the natural value, our species is operating as if the population were 30 times larger—nearly 250 billion people. It has been pointed out that when the world's population reaches 10 billion by 2100, the effective population would exceed one trillion.

These examples were modified from chapter 5 of Geoffrey West's book *Scale*. He sums up the point nicely by stating that, "...this exercise not only gives a sense of scale for how much energy we use but also illustrates how far out of ecological equilibrium we have come relative to the rest of the 'natural world.'" For me, a good reminder to be more mindful with the way I use my technology.

On finding motivation

I sat down this morning with the intention to write a short entry on supermassive black holes or some other enormous cosmic phenomenon, but simply could not find the inspiration or pin down a specific topic of interest. I've spent the last hour at my keyboard frustratingly writing a few incoherent sentences and then erasing them. Amid my impatience and lack of direction this morning, I was reminded that meaningful words cannot be forced out of the mind onto paper. Whether they are spoken or written, there must be some form of inspiration and purpose behind words for them to have any meaningful impact at all.

I stood up from my laptop with the intention of giving up trying to write for the day (a common occurrence and I usually succeed in my intent) and walked over to a window. Looking over the newly greened grass and budding trees of spring, I found inspiration! I realized that today was not a day to write about the coldness and vastness of intergalactic space, but rather a day to simply recognize and enjoy the beauty and warmth of our home planet. As soon as I had the realization, I sat back down and wrote this in five minutes. All that was needed was a little inspiration! The take home message for me: you cannot produce personally meaningful work without a proper motivation. When I force words out of my head, they find the page slowly and are incoherent. Once my mind has been lubricated by a proper motivation, words usually fall into place and I enjoy the process.

Doing things that are hard

Frustration. Uncertainty. Heartbreak. Failure. What would life be without these things? Not worth living in my opinion.

I've always found it amusing that the meaning of life is to be found in difficulty. And that despite this being recognized by the greatest thinkers for thousands of years, we are still deterred by things that are hard. It is paradoxical indeed, but it seems like we should learn: in order to grow and find meaning, we must subject ourselves to things that fundamentally suck. My homie Daniel stated it well when he said, "...to be unsettled, to want, and to struggle is the glory. White knuckled, bleary eyed, running yourself down, hitting everything hard. Making a mark." He went on to say, "...life can be a beautiful mosaic of unspeakably happy moments—and those mosaic tiles would be a meaningless blur if not for the grit between. While the show is still rolling...bliss and despair can be equally satisfying."

When times get tough—whether I'm stuck on a research question, reaching stagnancy on my writing, or having my heart broken—I appreciate my friends who remind me that *that* is what life is all about. I don't want you to tell me that it's just a wave of difficulty that will pass by; I want you to force me to embrace how that wave feels, right now. Remind me not to wait around idly for the dust of hard times to settle, but rather to actively explore the clouds that surround me.

So, I want to treat you the way I want to be treated. Embrace the chaos. If times are tough, feel them. Realize that those feelings, the low ones, are the contrast that allow you to feel the high ones.

Filling the room

Many of us enjoy the study of astronomy and other natural sciences because they serve as humbling reminders of our infinitesimally small place in the cosmos. One thing I find fascinating is how people react to discussions in astronomy that deal with vast cosmic distances and deep time. Many of us, myself included, seem to see things like this and think, "Oh my god, how wonderful" or, "Oh my god, how terrifying" (or more likely a combination of the two). Similarly, in discussions of deep time we may learn something that absolutely blows our mind—something that is awesome (in the real sense of the word). What I find amusing is how fast we humans are to move on from seeing comparisons like this or learning something about deep time. We have these momentary, identity-shaking moments of learning and discovery, and then we subsequently go back to worrying about what restaurant we want to visit for lunch. We learn that there are ~1 septillion stars in the known universe, and then concern ourselves with the way that people perceive us on social media. What the fu*#!

The things that human beings occupy themselves with truly does not make sense, but I suppose that is to be expected because of the staggering complexity of our own brains. Also, moments of awe are just that—moments. It is the brevity of amazement that makes it what it is. Additionally, it would not evolutionarily possible or efficient for us to be in constant terror of the vastness of the universe. Carl Jung discusses the fact that it is normal and necessary for us to "forget" things in this way to make room in our conscious for more important ideas and impressions. It is up to us to fill the room provided with experiences and ideas that are worth having.

Marks on a canvas

"Those soft bonds of love are indifferent to life and death. They hold through time so that yesterday's love is part of today's and the confidence in tomorrow's love is also part of today's. And when one dies, the memory lives in the other, and is warm and breathing. And when both die—I almost believe, rational as though I am—this somewhere it remains, indestructible and eternal, enriching all of the universe by the mere fact that it once existed."—Isaac Asimov, 1990.

What mark do you want to leave on this universe? It is easy to feel discouraged and unhopeful that any of us will make any sort of lasting impact on our community or the Earth, much less the universe. I find myself running into this occasional feeling of hopelessness. I have been frustrated and disturbed by the thought of my existence disappearing into space and becoming lost in the vast expanse of cosmic time. If you have not been confronted by such naggings, I would argue that you haven't attempted to face your existence on a deep enough level. Considering my lifespan on a cosmic timescale has been humbling to say the least, and has helped me frame how I want to spend my allotted time during this ridiculously strange and unlikely existence. Still, just when I think I have nailed down how I should live my life, thoughts of futility permeate my consciousness.

Asimov's passage provides beautiful insight into how to confront these thoughts, I think. It is reminder that your existence does not have to end when your life does—through love and memory, your existence will continue. Our bodies will be cold, but what we leave behind will remain warm. The world of rationality is a cold one, and I can't help but agree with Asimov in his hunch that our actions and creations will remain in some form eternally, beyond the existence of our species. Once we manifest something in this universe, it cannot be reversed. It becomes an intractable act in the play that is being played out on the cosmic stage. So it makes me think: what kind of lines do I want to leave behind? What mark do I want to leave on the canvas of existence? I genuinely hope it will be a mark of beauty.

External displacements vs. subjective experience

In the Malbec-fueled entry above I discussed legacy and the ease with which humans get tangled in webs of hopelessness when it comes to leaving a lasting impact on our surroundings. A swallowing of our goals, passions, desires, and accomplishments into the enormity of cosmic space and time seems inevitable. I've been looking for points of light shining through this backdrop.

One potentially comforting thought on this topic is that our existence does not necessarily end when our lives do. As Asimov eloquently stated, "...when one dies, the memory lives on in another, and is warm and breathing." Additionally, as far as we know, it is impossible to change the past. So, once we create something in this universe, it cannot be retracted. This is a thought that I think is not only uplifting, but existentially important. It emphasizes the importance of paying close attention to the actions you take and the words you say, because once they have manifested, they become indestructible components of the universe. But hold up. Why am I approaching this question as one that is fundamentally separate from my actual experience? Why am I focusing on the external effects of my actions, when perhaps the present moment inner experience of myself and others is what matters?

I came across a quote in the 1891 novel *Tess of the d'Urbervilles* that provided a refreshing perspective on these questions. It states that, "...the measure of your life depends not on your external displacement but your subjective experience.

If I am happy, and those that I love are happy, can that be enough? Because if so, excellent."

One of the main points I extract from this passage is that, whether we recognize it or not, we all have a legacy. Our mere existence makes it impossible not to. Perhaps I should focus less on my external displacement and pay more attention to the subjective details that give life glory and meaning. Love your family and friends, love yourself.

Empty sheets

I sat here looking at a blank document with many potential topics of discussion shuffling through my head: the bicameral mind theory, nature's indifference toward Homo sapiens, evolution of the cactus, earthquakes, the circulatory system. All worthy topics, but they were all distracting me from the beauty of what lied before me—a blank document. The blank document is a beautiful thing, you know. It is what allows us to start something new. Whether we are painting or writing or sketching, the unmarked paper offers an opportunity to do almost anything. Cliché I know, but this recognition got me thinking this morning about how the blank canvas relates to the mind.

The human mind is usually a goddamned mess. Ideologies, emotions, angst, desire, schedules, names, numbers, and random mindless chatter. If we consider the foundation of our consciousness to be an unmarked canvas like the one described above, our common experience is akin to dropping a bin of garbage atop it. It's all marked up and dirty and ugly. If you fall into a deep enough trap of thinking, you may not be able to see the clean white slate that once existed and might become convinced it never did. Perhaps a reason that many of us find it difficult to start something is because our canvas is compromised. How can you decide on what to create when there is no room for doodling? Must....Doodle.

Meditation offers a slow but sure cleaning up of your sheet, whatever form it takes. The multitude of thoughts and experiences that we identify with can be scrubbed away and provide us with glimpses back into when we existed with the ability to write and draw freely.

Now more than ever it is easy for our blank slate to be smeared with gunk and jumbled with words. Sitting down, closing the eyes, and focusing on the breath helps us realize this. And I think this simple realization is what will help clean it up. I'm grateful for the blank document.

What goes around comes around (as they say)

"Follow through on all your generous impulses. Do not question them."—Epictetus, 100 AD.

These were a few of the last words I read before leaving my apartment this morning, and they ended up playing an interesting role in how my day played out.

I arrived at the library on campus earlier than usual to read a paper prior to a class discussion. Relying heavily on caffeine in the morning, I wandered my way into line for the small library coffee shop. It was vacant except for a single person who was in an awkward situation of not being able to pay for his pastries and coffee due to the shops malfunctioning card reader. As soon as the situation registered, I remembered what I had read prior to leaving my apartment: never suppress a generous impulse! I certainly had one and felt no hesitation to offer to pay for the strangers' items. He resisted, saying it was over $12 (damn homie, running up a check at the coffee shop). I was committed and told him it was no problem. He was appreciative, told me his name was Michael, shook my hand, and walked away. I added my cup of black coffee to the bill.

As I left the library, I pondered the general idea of Karma. Karma refers to the primarily Hindu and Buddhist principle of cause and effect, and states that the future of an individual hinges on their previous actions. Broadly speaking, it refers to the fact that performing good deeds in the present will lead to good happenings in the future. What goes around comes around (as they say). I didn't really need to justify the action I took, but it was comforting to think that maybe someday my action of covering the dude's check would pay off. I got on with my day and it was a pleasant one. Around 5PM, I began my walk home.

I walked on icy sidewalks with my friend off campus and had a nice conversation. As we crossed a road, something nestled in the snow caught my eye. I detoured slightly and, much to my surprise, found the object to be a wet and crusty twenty-dollar bill. Thoughts of karma permeated my consciousness again, I picked up the bill, couldn't help but smile, and made my way home.

A cold winter

It gets cold in Montana. Damn, it gets cold. Each winter, ice hugs the roads in sheets as thick as the pavement itself. Snow of varying density stratifies up to a meter thick, smothering plants and stressing the branches of leafless trees. The feeling of having one's bare feet in the grass fragments into a memory. During these months, normally simple tasks such as driving, or even walking, can become difficult. One must dig out the vehicle, scrape the windshield, and warm the engine before venturing onto a normally wide city road that snow has narrowed to barely a trucks width. At a certain point, going through these motions becomes the norm. It is usually then that things begin to change.

The days begin to get longer; the Sun releases its daytime hug with the southern horizon. This in turn begins to peel away the layers of ice and snow. Dirty roads and dull yellow grass are slowly uncovered. Warmth creeps up the mountains and the rivers explode with sediment and debris-filled water. Gradually, and in amazing fashion, life returns. Those once naked trees bud (many flower) and are soon full of rustling leaves. The ice-encrusted plants of past months are alive! They explode too. The grass comes back with a vengeance and soon fills the spaces between our toes, piecing back together that memory that we almost forgot. A reminder that the darkest and coldest times are followed by some of the most beautiful ones.

A leading expert

Most of us fortunate enough to obtain a formal education learn some calculus between the ages of 16 and 24. This fact may seem uninteresting, but like many of its kind, it is quite remarkable.

Let's assume that you know at the very least basic derivatives and integrals. In today's relatively well-educated society, knowing these operations does not make you stand out. In fact, any employer in the competitive realm of science, technology, medicine, etc. assumes that you possess the ability to solve these types of problems. Your knowledge of single variable calculus is not impressive. Now let's take those skills and transport you back in time four centuries. You are now the leading expert in mathematics on planet Earth. Nice! This concept can be applied to essentially any other discipline (other than history, that would create problems, haha). Extract any modern undergraduate student in geosciences, biology, engineering, computer science, or astronomy, send them back 400 years, and they are the leading expert in their field. I was thinking about this fact this morning over a cup of coffee and found it entertaining.

Another thing—consider something that you are mediocre at, or maybe something that at the surface seems unimportant or unimpressive. For example, you had the ability to drive a giant metal machine to work this morning. Filled with dials, screens, sticks, pedals, and mirrors, your everyday vehicle would petrify every person on Earth only 200 years ago. You mindlessly hop in and send it! Maybe I'm trying to make you feel better about yourself, but it is objectively impressive what all of us can do. We are the individuals living through the inflection point of an exponential change in humanity and it is *our* brains that are expected to adapt to these changes. We have to familiarize ourselves with the collective work of our predecessors simply to build a foundation for new insight and discovery. So, don't be down on yourself today. Pretend it's 1650 and let the confidence flow.

Love and the present moment

"...you can throw yourself flat on the ground, stretched out upon Mother Earth, with the certain conviction that you are one with her and she with you. You are as firmly established, as invulnerable as she, indeed a thousand times firmer and more invulnerable. As surely she will engulf you tomorrow, so surely will she bring you forth anew to new striving and suffering. And not merely 'some day': now, today, every day she is bringing you forth, not once but thousands upon thousands of times, just as every day she engulfs you a thousand times over. For eternally and always there is only now, one and the same now; the present is the only thing that has no end." -Erwin Schrödinger (1951)

Ruminating on the past, anxiously awaiting the future; this is how we spend most of our time. What a silly way to live. As Schrödinger articulates above, these moments do not even exist. What exists is right now—we are entirely consumed and surrounded by the present moment, yet we don't recognize it. I think we all do get brief glances of the present moment. It fleetingly roars in our ears and enters our eyes; we recognize its brilliance, but it happens too fast to appreciate. Like light reflecting off a moving surface, our recognition of the present comes and goes too quickly to comprehend. I've wondered why it isn't easier to experience. Why doesn't it stick around, why does it slip through my mind like sand through my fingers? Is it too powerful?

Contemplate the moments in which you are not concerned about the past or the future? What are you doing? Maybe it's found in work that you enjoy, solving problems perhaps. That's fine. Maybe it is found in love. Love. What a sick and wonderful thing. When you look into the eyes of a person you love, the present is found. Fully captivated, fully engaged, entirely relaxed. The present moment can be equally found in the loss of love. When you're able to shelf regrets and memories, there you sit with nothing but feeling. Sadness and loss. These feelings are real and real powerful; both sides of love can engulf you in the present moment. Another reason it rests at the top of human emotion.

A bound existence

Why do we have an existence so limited? So bound and constrained. It occasionally gets on my nerves. Yes, our lives are staggeringly unlikely, and to live in any state other than amazement is ungrateful. But I find my mind drawn to bigger, grander places.

Sometimes when I close my eyes before sleep, I imagine myself with fantastic, unconstrained abilities. The ability to fly and see the world from above, then dive into the depths of the sea and see it from below. I dream of seeing our blue planet from just beyond the moon, then as a white speck from beyond Mars. I grasp at what it may feel like to float through the interstellar and intergalactic voids of the cosmos. What does it feel like to swim through a vacant immensity of nothingness (shoutout H.G. Wells)? I wonder. I contemplate other universes with different physical constants, different forms, different dimensions. I think about what love means in these far places. I bring myself back to Earth and consider my longing for communication with other creatures. I leave again and am consumed by a wormhole in the outer reaches of our universe; I visualize myself as a hero and as a loser in epic battles on its other side. I ponder what it would be like to see the entire universe at once, with no superposition of sights, sounds, and feelings. I come back into my head and acknowledge that it was all in my imagination. I'm frustrated because I want to go to these places—not in my head but in real life.

But the question arises: what separates my imagination from reality? Perhaps what I am seeking is right behind my eyes.

No other way

"Love is such a mammalian emotion...and I'm glad to be an animal. I'm glad to be side-by-side with all of you. Because happiness isn't real unless it is shared. And whatever this is, if we're emerging from nothingness and descending to nothingness only to enjoy a few crystallized moments of this existence, I'm glad I didn't have to do it alone." –Daniel Fenstermaker

We are in this together and it is going to kill us all. Regardless of the position you occupy in this world, whether basking in the light of success or recovering from an impact with rock bottom, we share the same fate. Consciously aware of this fate, I've always wondered what keeps us going. I've reframed the question and concluded that I would not want it any other way. As my homie says above, "...I'm glad to be an animal." I would not trade my complex and conflicting emotions for anything—they are what make my life worth living. Nothing builds my character more than working through a difficult emotional situation. These situations remind me of how unpredictable and uncontrollable the cosmos is, and the fact that each living human being has them brings us together in an unparalleled fashion. Indeed, much of the work that good people are doing attempts to maximize the well-being for the greatest amount of people.

One other thing on human emotions—there is nothing as real. Even if we live in complete delusion as to what the universe actually is and why it is here, you cannot argue that I do not have conscious experience. And you cannot take from me what I have and am currently experiencing. Love lies at the pinnacle of powerful emotion. It fuels extreme effort, fulfills many individuals search for meaning, and without a doubt has the potential to shred the mind and body if it backfires. Love is a mammalian emotion—it will continue to break us down and build us up in a most painful fashion. I would have it no other way.

Dukkha

During the first year of my undergraduate, I read *The Dhammapada*, which is one of the most widely read of all Buddhist scriptures. I think that much of my interest in philosophy sprouted from reading this text, as it opened my eyes to philosophy of mind and how it is possible for one to be in control of one's emotions (or at least in control of how we respond to our emotions). Over the past five years or so, I have read many other great Buddhist texts that have helped me to various degrees, but I have always been torn by their central message. My homie Daniel (quoted in the previous essay) wrote a passage during the summer of 2019 that mirrored my thoughts on Buddhism almost exactly. He begins his essay by stating that the core of Buddhism (generally curing oneself of desire) has always passed him by and follows with the following explanation:

In Buddhism, the only cure [to suffering] is to rid yourself of desire. To sit idly by in a state of pure contentment (which is nice) but also a state of inaction. I do not abide. Sure, the kill is never as rewarding as hoped or imagined. But the hunt is the fun part. To be unsettled, to want, to struggle is the glory. White knuckled, bleary eyed, running yourself down, hitting everything hard. Making a mark. I believe I have one life and I'm not going to meditate my way through it...dissolving my individuality.

Shatter the ground when you hit it, love so hard that it hurts, chase your prey until your body breaks. The highest highs are worth the lowest lows...additionally, I postulate the highest highs are unachievable without the lows. I don't think true love is steady-state contentment. I think it is a cycle of friction and resolution. Life can be a beautiful mosaic of unspeakably happy moments—and those mosaic tiles would be a meaningless blur if not for the grit between. I think sometimes you need to be freezing cold, in the muck, to really feel

*those hot showers. I don't think that you should meld into the
background of life and become a whisper on the breath of the breeze.
While there is still sand trickling in my hourglass I will not rest. I
suspect we all get peace in the end. While the show is still rolling,
though, bliss and despair can be equally satisfying.*

A friend indeed

*"What progress have I made? I am beginning to be my own friend.
That is progress indeed. Such a person will never be alone, and you
may be sure he is a friend of all."* –Seneca, 50 BC.

In *Letters from a Stoic*, Seneca the Younger
emphasizes to his companion Lucilius the importance of being
able to pass time by oneself. I agree with Seneca's claim that
nothing is a better proof of a well-ordered mind than an ability
to stop and exist alone. This is especially relevant today
considering that the 21st century is one of increasing distraction
(we are the first people ever who need to learn to find a balance
between the positives and negatives of our technology). Part of
a healthy balance, I think, consists of occasional total isolation
from these tools. If you are unable to go a day without your
phone, it can no longer be considered a tool—it is a vice. Soon,
it is likely there will be no distinction between us and our
technology. Until then, isn't it desirable for one to immerse
themself in the natural world? It seems to me that this
immersion can be amplified and more meaningful if it is done
alone. Whether we are playing our favorite sport, reading,
watching a movie, or sitting in meditation, time alone is
invaluable to our continued mental and physical health.

Being with people is tight too haha. My relationships
and community and friendship really hold me together. So, it's
a balance (duh). Occasional isolation from civilization has the
potential to give one great insight, but nothing is worth having
unless it can be shared.

Adrift in the void

"Few people without a training in science can realize the huge isolation of the solar system. The sun with its specks of planets, its dust of planetoids, and its impalpable comets, swims in a vacant immensity that almost defeats the imagination. Beyond the orbit of Neptune there is space, vacant so far as human observation has penetrated, without warmth or light or sound, blank emptiness, for twenty million times a million miles. That is the smallest estimate of the distance to be traversed before the very nearest of the stars is attained." -H.G. Wells, 1897.

It can be easy to forget that we exist in isolation, adrift in a sea of nothingness. Earth's atmosphere provides us with the familiar, deceiving comfort of blue skies during the daytime, when in reality our planet is one alone and enveloped in darkness. As made clear in the passage above, there is almost nothing filling the void between our outer solar system and the nearest star, Proxima Centauri, is loitering 4.24 light years from Earth. What may sound like a relatively short distance to travel is disturbingly far. It takes light waves—traveling at the absurd cosmic speed limit of 1 billion km/hr—over four years to reach us from Proxima Centauri. Of course, even our fastest spacecraft do not come close to reaching the speed of light—one of our fastest craft, the New Horizons probe, travels at a mere ˜60,000 km/hr. Assuming a straight path at this speed, it would take us over 80,000 years to reach this nearest star (translating to approximately 2,700 human generations). Between us and Proxima Centauri exists nothing but the vacuum of space, with an average of one hydrogen atom per cubic centimeter.

Our solar system, floating in an obscure corner of an average galaxy that is 100,000 light years across, exists as a mote of dust. There is no sign of any other cosmic inhabitants other than us, and it is possible that we exist entirely alone in this universe. I am thankful that we have each other.

Arbitrary time

Time. A great deal of angst arises in human life because of the concept of time. Rumination on past times leads some to depression and uncertainty regarding future times manifests as anxiety. In Part III, we will discuss how taking an astrophysical/general relativistic view on time can help mitigate feelings of time-related angst. For now, it is sufficient to say that time is not absolute (it varies relative to the strength of an observer's gravitational field). Not only is time dependent on external variables, but our human descriptions of time are arbitrarily named and arise because of natural phenomena that are out of our control. For example, we assign the word "day" to the time it takes for our host planet to make one rotation about its axis. We break this 24-hour period into "day" and "night" based off when the sun dips beneath and rises above the horizon. Our descriptions of time are specialized to the planet we live on. Our essentially meaningless time classifications become apparent when we ponder them from a galactic or cosmic scale. There is no "day" in space, no "month", no "year". They are words specialized to our miniscule place in space.

How does coming to this realization help mitigate feelings of angst? Think about how often we worry about something that is going to happen in "a few days". Or consider how many times we have concerned ourselves with occurrences that happened days, months, or years prior. I think recognizing that these are simply words we have created can make it easier to let go of them. Not only that, it also helps in understanding that those past and future moments do not actually exist. They are concepts, ideas, and thoughts. When tomorrow arrives, it will be the present, and only then does it exist.

Scales of observation

We look at images of deep space and of the microscopic world in complete awe. It seems to be much more difficult to view the world at our normal, human scale of observation and be amazed. It's simply because we are used to it. As I've said in a previous entry, repetition breeds indifference. The scale at which we see things is truly as interesting as those observed by the Hubble Space Telescope or a scanning electron microscope. We simply need to break the spell of thinking that our surroundings are "normal", and curate mindfulness that allows us to acknowledge the true magnificence of the scale at which we live.

Isn't our scale of observation by definition as interesting objectively as the smaller and larger scales that surround us? The only reason we are not in frozen awe every waking moment by its beauty is because we have all been observing the world at this scale for our whole lives. This is why we cherish the opportunity to look down through microscopes or up through telescopes—it changes our perspective on the universe. With enough practice, we can recognize that our indifference towards the world around us is simply a result of the time spent at our scale of observation. After recognizing this, it will not be necessary for us to change our scale in order to be amazed. We have the power to change our perspective on the universe using nothing but our current senses. Today, take a few moments and apply mindfulness to the task at hand, and consider that the scale at which you're working is not as normal as you think it is. I will join you.

Life on a cosmic scale

Are you impressed or depressed by the minuscule amount of cosmic space that we occupy? Or perhaps it evokes other feelings? I have reformulated my own approach to this question and am developing a new perspective.

In the past, I have taken an optimistic stance regarding the tiny rock that we inhabit. "How fortunate we are to occupy such an unlikely position in the cosmos!", I have said. My optimism has been countered by people who think that our immense smallness serves as justification for nihilism. "How insignificant and limited we are! Our goals and accomplishments will ultimately be enveloped by the vastness of space and forgotten forever," one may say. You must admit that both sides have a case to make. Indeed, both sides are probably right! I have been pondering this this argument, trying to do it without considering the Earth (and its inhabitants) as entities separate from the rest of the universe.

Alan Watts writes extensively in support of the argument that the universe essentially doesn't exist without us in it (recall the work of Donald Hoffman). Watts and many others emphasize that without an entity to consciously experience it, the universe does not exist (at least how we know it). In other words, we require the universe, and the universe requires us. One may argue that we do not rely on one another, but that we ARE one another (i.e., we are the universe expressing itself). So, revisiting the question: should we interpret our small place in space as saddening or enlightening? If you buy the idea that we are not separate from the universe, this isn't even an answerable question. It is analogous to asking, "Is the size of one water molecule in the ocean depressing?" Of course, the answer is no, because we know that the water molecule is one of a huge number of molecules that contribute to create a deep and restless and powerful ocean. A recognition that we are not separate from the universe blows up the original question. At the very least, there is nothing depressing about contributing to something far greater than ourselves.

Are we the universe?

Philosophers have wondered if we were created by the universe so it could be conscious of itself.

I came across a thought-provoking quote from Alan Watts the other day that reminded me of the consideration above. He states that, "...through our eyes, the universe is perceiving itself. Through our ears, the universe is listening to its harmonies. We are the witnesses through which the universe becomes conscious of its glory, of its magnificence." Carl Sagan famously stated something similar: "We are the way for the cosmos to know itself."

I think that this idea could help us answer the question of why consciousness exists in the first place (the hard problem of consciousness—why is it like something to have an experience at all? More on this later). If we were created by the universe so it could experience itself, we have our answer, right? The universe is using us to learn about itself—we experience so that it can experience. Collective consciousness. I'm not entirely sure what I mean by "it" here, but I suppose "it" in this context takes us into the territory of *pantheism*. Pantheism is a religious doctrine that considers God not as an entity separate from the universe, but rather God as the universe. From this perspective, everything in the universe—a flower, your lover, a snowflake, the sound of a trumpet—are constituents of God. No separate creator. So perhaps we can answer the question of cosmic genesis with our initial idea haha. This idea can be compared with discussions on panpsychism—the theory of consciousness which generally states that consciousness is a ubiquitous and fundamental property of matter itself (pan, "every, all"; psych, "mind, spirit"). They are not the same concept, but both play with how and why conscious experience arises in the universe. The biggest questions of our existence.

Taking things for granted

Why is it so easy to take things for granted? The question has nagged me for years. With such a short time to experience and such uncertainty oozing through the objects surrounding us, why is it that we remain tangled up in unmindful thinking and tripping over negative emotion? It is paradoxical. My goal each day I rise is to do my best to clear these webs and consciously recognize the beauty of each moment. To acknowledge the smallest details; the angle of the sunlight hitting my desk, the smell of garlic while I was cooking lunch, the feeling of fabric against my skin. Although these moments of mindfulness will be the ones that I cherish when the end arrives, I mostly take them for granted. And it is easy to do so; it's the default! One of the fundamental reasons I think this is the case is because these moments occur so frequently— we become indifferent, regardless of how magical an experience is. Along a similar track, we take things for granted that seem everlasting—things like our health, family, and life in general. The steady and dependable presence of certain things begin to blend into the backdrop of our lives. Of course, impermanence is a fundamental truth of existence, so we are setting ourselves up for angst and disappointment when things inevitably change.

So, what can I do? I think is necessary to train myself to pay attention to what is important—to be captured by the things that if this day were my last, I would be happy to have given my attention to. There is no better time than right now to acknowledge the beauty of the existence that the cosmos has presented us with. Sometime today, wherever you are, take a moment to feel gratitude for something you usually take for granted or look past. As always, I will join you.

The traffic light

This morning, alone on the way to work, I found myself laughing out loud as I stepped off the sidewalk to cross the street. Walking on one of the busiest streets in my city, I was stopped waiting for the little fluorescent man to appear, signaling that it was safe to cross. Traffic was bombing by in front of me; I recall a massive and dilapidated truck bouncing by labeled "Gary's Towing" that seemed to be totally out of control. Meanwhile, a woman in the passenger seat of a car waiting for the red light to turn was yelling at her unfortunate driver about something. I couldn't help but look over at her; cigarette in a tense hand, she locked eyes with me and yelled, "What the fuc* are you looking at?!" My gaze went up and across to the adjacent traffic light, I cracked a smile. Chaos on all sides.

A few seconds before the light changed, I was overwhelmed by a feeling of wonder. I was contemplating how much I take for granted in the developed world. Things like traffic lights. I was surrounded by a bunch of relatively smart primates in metal contraptions, all of whom were waiting for a light to change from red to green. Waiting for "their turn" to go. What a ridiculous idea! Yet, almost everyone obeys. There are systems put in place to make our lives easier and safer that we usually do not express one sliver of gratitude for. I highly doubt the fuming lady with the cigarette was thinking about how convenient traffic lights are.

It seemed like a strange time and place to be grateful, but I was. One of my primary goals in life is to be mindful of things like this. To not let the repeated occurrence of something lead to a disregarding of it. I stepped off the curb laughing—that exact moment, surrounded by noise and tension, was the highlight of my day.

Captured attention

Commonly, the joy I experience during a moment of mindfulness is inversely proportional to its, let's say, societal significance. My favorite moments do not come from my work or from within the pursuit of a larger goal. Or even from my relationships with other people. They come when my attention is captured by the beauty and complexity of an unsuspecting object. The sound of leaf edges scraping in the wind. The sight of iridescent swirls of gas in a street puddle. The feeling of air passing through my hand as I walk. Sensing these types of things is a privilege and I try not to take them for granted. Such experiences are hard to write about because it is all recollection. True mindfulness consists of giving yourself to the present moment—consequently, you are not thinking about how you will communicate the experience later. Translating the sensation into words is the task of better writers than myself.

On the topic of mindfulness, Sam Harris writes that, "...there is no such thing as a boring object of attention. Boredom is simply a lack of attention." I love this point—it reminds me that every waking moment of the day offers the opportunity to practice mindfulness. The most boring aspects of being a human, like waiting in traffic or being stuck on hold, can be meaningful and exciting. Harris goes on to state that, "Concentration is intrinsically pleasant."

If today were your last day on Earth, would you be happy with the kinds of things that captured your attention?

Replaying the tape

If we were to rewind the tape of life a billion years, would evolution by natural selection inevitably lead to the existence of Homo sapiens? This is one of my favorite questions in all of science and philosophy.

In his book *Wonderful Life*, Stephen Jay Gould claims that if we were to turn back the clock, history would not repeat itself and evolution would have taken an entirely different course—resulting in an unfamiliar world almost certainly lacking human beings. His argument is essentially that the outcome of evolutionary processes depends too much on random events to create the same thing twice. This is fundamentally an indeterministic view of the cosmos, one that claims that stochasticity plays a key role in how the natural world evolves.

Determinism states the opposite—that all events are determined entirely by previously existing causes. From a staunchly deterministic view, nothing is random—every event that occurs is inevitable given its dependence on initial conditions. This includes events such as asteroid impacts, disease, etc. that have played fundamental roles in the evolution of life on Earth. Therefore, if we were to replay the tape of life, we would indeed see the inevitable emergence of humans. We are on tracks. It is important to emphasize determinism's sensitive dependence on initial conditions. If one air molecule were misplaced as the rewound tape began to play, it is possible that Earth history would take an entirely different trajectory.

Observe then interpret

"Only when armed with a full understanding of each step in the analytical process will researchers have a full appreciation of the strengths and limitations of their data."

This quote from geologist Dr. George Gehrels serves as a reminder to obtain a thorough understanding of how something works before you begin using it to solve problems. Although his quote is referring to a specific analytical technique within the geosciences, the primary message can be extrapolated to essentially everything in life.

With specific regard to technical application and data acquisition, there are many reasons why a deep understanding of your techniques must be obtained. I think one of the most important reasons is to avoid misinterpreting corrupt data. If one does not understand the ways by which data can become corrupted, one cannot recognize it in the results. This would obviously have huge negative implications on interpretations of a given data set.

Expanding our scope, we will think about general problem solving in everyday life. In analogy to George's quote, here we think of the "analytical process" as life's happenings and "data" to be the current state of our problem. When attempting to solve a problem, I know that quickly jumping to conclusions without thinking through the possible mechanisms by which it arose is the worst strategy I can take. If I do not consider how a problem in life has arisen, how am I to approach solving it? I must "arm myself" first with an understanding of how their current problem manifested itself, then begin the process of interpreting it in context of my current life and ultimately tackling it. Just as I cannot recognize corrupt data in scientific results without an understanding of how they can arise, I don't think I can recognize why we face our current problems without an acknowledgement of their root causes.

The lessons of the scientific method go far beyond the laboratory.

Mortal

Death has been an unsettling thought for me, and I think a lot of my interest in philosophy has been fueled by a desire to understand why my friends, family, and I must depart one another in the not-so-distant future.

Stoic philosophy has provided me with helpful wisdom on the topic. In his collection of essays *Meditations*, Marcus Aurelius writes extensively on how to approach the end of life. He spends a great deal of time pondering the inevitability of death and articulates the fact that the longest and shortest lives come to the same end. He also addresses something that I've always struggled with: that people are afraid of dying because they feel like they are being deprived of potentially positive events in the future. He counters this argument by emphasizing that *you cannot be deprived of something that you do not have.* The future does not exist until it becomes the present moment, so with death, you are only being deprived of the present. This is one of many realizations that has helped me come to terms with dying. It also leads to a far greater appreciation of the present moment. It is all that we have, and we obviously will not get to experience it forever. So, make sure that you occasionally forget about your goals and aspirations and allow yourself to be carried downstream by the stream of life.

Scared of eternal darkness

"When I consider the short duration of my life, swallowed up in the eternity before and after, the little space which I fill, and even can see, engulfed in the infinite immensity of spaces of which I am ignorant, and which know me not, I am frightened, and am astonished at being here rather than there; for there is no reason why here rather than there, why now rather than then. Who has put me here? By whose order and direction have this place and time been allotted to me? . . . The eternal silence of these infinite spaces frightens me." –Blaise Pascal, 17th century mathematician.

It seems like our lives might be bound on either side by infinite nonexistence. For me, and perhaps many others, it is difficult to imagine what this even means. What does it mean for time to go on forever? What does it feel like to be dead forever?

I contemplate my own mortality and I've learned something interesting about myself in the process. Sometimes the thought of my own death and the death of my friends and family disturbs and frightens me, and other times, I am overwhelmed with feelings of gratitude and freedom. Why the dichotomy? And which feeling should I embrace? I've faced a fear of my inability to grasp infinity and clung to the shelf break that drops into an abyss of meaninglessness, but I've also faced a cosmos full of uncertainty with square shoulders and been brought to tears by the beauty of life. And through these feelings, it seems like the appropriate way to react to our existence is the latter.

Our existence seems infinitely unlikely. Therefore, it seems like an awful waste to live in any state other than amazement and appreciation. We can love, suffer, see, hear, touch, and think—these things by themselves are worth living for. For all we know, we are the only organism in the entire cosmos, billions of light-years wide, that can experience these things. We should not take it for granted. Additionally, there is no reason to fear eternity, for eternity is only frightening if we are there. And I don't think we will be, at least not with our current form of consciousness. After we die, we will probably return to where we were for the billions of years before we were born, a place where nothing exists—no time, no feeling. The eternal silence that awaits me will indeed be as if I never woke from a wonderful slumber. The ultimate state of tranquility. But until then, I will be grateful for the time that I have and strive to create my own meaning.

How much time do I have left?

    I have regurgitated the argument that the dramatic shortness of human life is what gives it beauty and meaning. I still lean is this direction, but damn, when it is put into perspective, life seems too short.

    Recently I was pondering the probable percentage of my life that has already passed. If I am fortunate enough to live to the average age in the United States (~80 years), I am 30% there! This value may not seem that high by itself but viewing it through a different lens changed my perspective. The figure below represents an interesting way to visualize the amount of time that each of us may have left to utilize and enjoy (inspired by Tim Urban's blog, *Wait But Why*). When I looked at the boxes that I've filled in and considered how fast they seem to be going recently, this served as a humbling reminder that life is

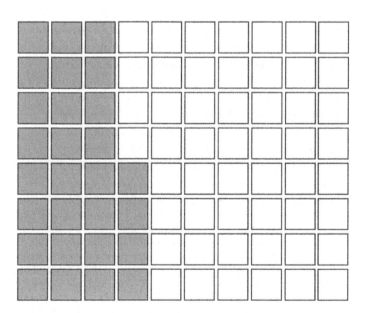

going to absolutely fly by[.]. I was talking about this with a close friend over beers, acknowledging the fact that before we know it, we will be in the latter years of our lives. This visualization technique can be adapted to filling in boxes that represent months, which is arguably more unsettling considering how fast months can go by. Although these experiments can be discouraging and intimidating, I think they can also be inspiring and motivational. If I consider my life to be a book (sorry for the cliché), I am rapidly approaching the middle. Knowing this, it is time for me to think about—if I were reading it—what I would want to happen.

A brief existence, indeed. Read great books, climb tall mountains, drink great wine, be kind!

Why?

I believe that the question "why am I here?" arises mostly out of the delusion that our species is separate from the rest of nature. When you consider us for what we are—a part of nature—the question begins to lose its sense. For example, let's consider liquid water (which we usually think of as nothing more than a part of the natural world). We are not constantly asking the question "why is water here?" and we don't question its existential meaning. We simply know and accept that it exists and are not affected by whether it has an ultimate universal meaning. Of course, we are justified in our wonderment towards why anything exists at all, but it is beside

---

[.] I was putting the finishing touches on this book and realized that the box figure needed to be updated. Only 25 boxes were filled in. I finally finished this book at age 28—it was depressing but also empowering that one of my last actions before publication was to fill in three more boxes. Also, the figure made me think of how uncertain and unpromised those empty boxes are. These are the kinds of thought experiments that give me the courage to share this kind of a book with you – one that never really feels like it's done and contains a lot of writing from my earliest contemplations on what life is. Life is going to go by fast and I know I would regret if I held onto these reflections just because it felt vulnerable to put them out there.

the point. I think a healthy approach towards our own existence would mirror that which we apply to other parts of nature. Like a tree or a river, we do not exist to fulfill a destiny of universal importance. Our existence grew out of the laws of nature, and the laws of nature will ultimately decide where we end up. We must remember that nature does not know that we divide her into separate parts.

The origin of consciousness

The problems surrounding the brain and the origin of consciousness are as interesting as any in the known universe. The complexities of the brain surpass those found in any of the other sciences, and I believe that the most fundamental questions surrounding human existence will be answered through continued neuroscience pursuits.

The brain is everything it seems. Without it, there is no consciousness. Obvious at the surface, this is a statement that flies in the face of centuries of philosophical discourse regarding a metaphysical "soul" that is separate from the assemblage of skin and blood and guts that we call our bodies (←haha this sentence made me laugh so I kept it in). Modern neuroscience research continues to find no evidence for a soul. On the contrary, it seems like consciousness is a result of the physical mechanisms of our brains operating under the laws of physics. A piece of evidence that I find particularly supportive of this interpretation is regarding brain injuries. Our mushy, wet brains are vulnerable to all different kinds of malfunction, and different areas of the brain can be injured or effected by various health problems. When certain areas of the brain are affected, human beings lose their abilities to do or recognize certain things (e.g., playing the piano, seeing color, etc.). With an ever-increasing knowledge of the role that different areas of the brain play in human ability, we are uncovering the fact that how we experience the world is a consequence of the entire brain working as an interconnected organ. If an eternal soul separate from our physical bodies were responsible for our consciousness, one would not expect a knock of the head to strip us of certain conscious abilities. This is one line of reasoning that supports consciousness being a manifestation of biology.

Quantum consciousness

Quantum mechanics (QM) states that there is something profoundly fundamental about how we observe or measure a system (i.e., the state of a system drastically changes when we look at it). One example is how an electron orbits the nucleus of an atom unobserved (in a cloud, or wave function) and what we see when we observe it (as a particle, with a position in space). This concept, known as wave-particle duality, is troubling for physicists and neuroscientists who are attempting to unravel the mysteries surrounding human consciousness. The problem is this: if QM defines fundamental properties of the cosmos, why should it require a conscious observer? Some argue that in a universe governed by QM, conscious beings (along with other macroscopic objects) should emerge as a natural consequence, not the other way around. It is conundrums like this, alongside others such as the Heisenberg uncertainty principle and quantum entanglement, that bring into question whether consciousness can be explained by QM.

The cosmologist Roger Penrose proposes a playful and much criticized quantum gravity theory of consciousness in his 1989 book *The Emperor's New Mind*. The theory is highly speculative and involves abductive reasoning from math that I don't understand (if you're interested, look up Godel's incompleteness theorems). Math aside, the theory posits that QM is significantly involved in brain activity, and therefore, consciousness. More recent studies "...add biological flesh to the quantum mechanical bones..." of Penrose's theory, as the neuroscientist Christoph Koch would say (see Hameroff and Penrose, 2014; Simon, 2018). Koch doesn't think that such exotic physics is needed to understand consciousness. As he puts it: "It is likely that a knowledge of biochemistry and classical electrodynamics is sufficient to understand how electrical activity across vast coalitions of neocortical neurons constitute any one experience." He goes on to say that his mind is open to any plausible description of how consciousness arises. It goes to show that even the leading experts in the science of consciousness have no idea what it is and how it works.

What does it feel like to be you?

It feels like something! The feeling you are currently experiencing is the most fundamental and digestable definition of consciousness. The philosopher Thomas Nagel first proposed this thought experiment in an essay titled *What is it like to be a bat?* (Nagel, 1974). He (and many others) maintain the argument that an organism is conscious *if there is something that it is like to be that organism.* Following this line of reasoning, we can make a few assumptions with nearly 100% confidence, including that the chair you sit on or the shoes you stand in are not conscious (for a counter to this claim, see the following essay on panpsychism). It seems likely that a brain (or something like a brain) is a precursor to consciousness, because the felt presence of conscious experience depends entirely on the brains ability to send and receive signals.

I've been thinking a lot about where the line should be drawn in the discussion regarding what is and is not conscious. I find myself making serious adjustments to the line's position in the natural world. If we can conclude that a certain amount neurological complexity is all that is needed for "the lights to come on", it becomes likely that we share our planet with much more conscious life than we currently perceive. The bumble bee, for example, has over a million neurons within their brain that weighs <1 gram (10x higher circuit density than human beings; Li et al., 2017). Capable of highly complex behavior, bees have the ability to recognize individual faces within their nests and even dance with one another. With behaviors like this and vastly complex brains, I truly believe that an inability to communicate with bees may be the only barrier preventing us from recognizing their consciousness. This can be expanded to many other animal species and possibly even plants and fungi (some of which seem to communicate with us when ingested).

The origin and mechanisms of consciousness are fascinating, and I am mad excited to see where the next few decades of neuroscience research take us.

Panpsychism

An emergent phenomenon is a system that has properties greater than the sum of its parts. In other words, an object is considered emergent when it is observed to have properties that its constituents do not have by themselves. Perhaps the most obvious and important example of this is biological life, which is an extremely complex operating system that emerges from much simpler properties of chemistry. Snowflakes are one of my favorite examples, representing another impressive example of emergence in a natural system. Emergent behavior arises due to (often simple and well-understood) interactions of constituent parts with each other and the environment. Interestingly, there appears to be no clear "leader" that directs the behavior of an emergent system—they simply manifest in a way that is dependent on initial conditions. I define emergence not to discuss biological life or snowflakes, but instead the far more enigmatic phenomenon of consciousness.

Consciousness, or experience, has been explained by many scientists and philosophers as an emergent phenomenon that arises solely as a result of neurophysiological processes. They make the physicalist argument that just as life arises from complex chemical reactions, consciousness arises from physics in the brain. This theory of consciousness as an emergent phenomenon locks horns with the recently reignited theory of consciousness known as *panpsychism* (Nagel, 1980).

Broadly speaking, panpsychism states that consciousness is a ubiquitous and fundamental property of matter (pan, "every, all"; psych, "mind, spirit"). Although it can take on a variety of forms, I prefer the definition that entertains the idea of all matter being imbued with consciousness (a view criticized as being untestable). Panpsychism is incompatible with emergentism because it states that consciousness—rather than emerging from physical brain processes—is built into matter itself.

Panpsychism has gained traction as philosophers and neuroscientists have explored the *hard problem of consciousness*, which emphasizes that even if we completely understand the physical processes in the brain that lead to consciousness, we still cannot explain why it feels like something to have subjective experience (e.g. why does a certain experience accompany tasting basil, or smelling garlic, or touching velvet?). Thomas Nagel makes a "non-emergence" argument that supports a panpsychic view by stating that, "...mental properties cannot be reduced to physical properties" and that "...all properties of complex systems derive from the properties of its constituents" (Nagel, 1980).

Technicalities aside, it is fascinating to entertain the idea that everything material, however small, has an element of individual consciousness. Since we don't have even a remote idea of how and why consciousness exists, we need to entertain all possible explanations. So, when you're going about your day, I encourage you to consider the idea that everything surrounding you may be infused with consciousness and may even be aware of your presence.

Expanding on the hard problem

The *neural correlates of consciousness* (NCC) are defined as the minimal neuronal mechanisms jointly sufficient for any one specific conscious percept (Chalmers, 2010). Put more simply, what biophysical processes take us from a glob of grey goo behind the eyes to a world of vivid experience? We can probe such questions empirically by monitoring neural changes during specific subjective experiences. For example, which part of the brain lights up when you see a flower, hear a saxophone, or taste red wine (and at what intensity, and for how long, etc.)? Studying such experiences has uncovered that specific neuronal changes do indeed correlate with specific experiences. A major contribution in the long-lived debate known as the mind–body problem, which concerns itself with the relationship between consciousness in the mind and physical processes in the body.

Another related debate exists, one that is perhaps much stickier: the hard problem of consciousness. Formulated by the philosopher David Chalmers, this problem confronts a deeper issue nestled within the NCC: even if we can describe the physical processes in the brain that integrate and categorize information, it still does not answer why these processes are accompanied by experience. For example, a certain firing of neurons occurs when I see a sunset, but we cannot know WHY the firing of those neurons generates the specific experience it does. Why not a different experience? Another example is why certain nerve stimulation causes pain rather than pleasure, and vice versa.

A thought experiment considers a human super-scientist who is locked in a black-and-white room with limitless logical acumen and learning material on the physical world. The person learns the entirety of all physics, chemistry, biology, etc., eventually leaving the room and experiencing color for the first time. Jackson (1982) points out that even though the person has a perfect knowledge of the physical world, they must uncover new truths about what it is like to see in color once leaving the room. There are absolute truths about consciousness that cannot be deduced from a limitless knowledge of the physical.

It is like something to experience color, and we don't know why. Many scientists and philosophers consider the hard problem of consciousness to be unsolvable.

Veil of thought

Usually without realizing it, we spend a disproportionate amount of our lives viewing the world through a veil of irrelevant thought. This is one of the truths that is revealed when you start a consistent meditation practice.

We live in a world of increasing technological and social complexity, and with these changes come more distractions from the roots of true human experience. When you are distracted by the internet or non-mindful conversation, it is almost impossible to recognize what your mind is doing in the background. When I sit down alone in a quiet room or outside with no distractions is when I recognize what is running in the background. That's when I realize that the mind is similar to a laptop with hundreds of minimized programs that are still running full speed. Most of us know all-to-well the effect this has on a computer, yet we are oblivious to the effect it has on our minds and bodies. These background programs are the irrelevant thoughts that cloud our perception and overall experience of the external world. They slow us down and cause unnecessary harm and angst. Mindfulness meditation (with enough practice) allows one to recognize the programs running in the back of the mind and close them (or at least organize and better understand them).

With one 10-15-minute meditation session, I am abruptly exposed to how frequently thoughts arise and how significantly they shape my perception. With practice, it becomes easier to let these thoughts pass through and ultimately gain insight into the true nature of existence.

A floating brain

In Annaka Harris' book *Conscious: A Brief Guide to the Fundamental Mystery of the Mind,* she articulates a helpful thought experiment regarding our subjective experience.

She first has the reader imagine a fresh brain without any sense organs connected, floating in empty space. Then she has us consider what it would be like to turn the sense organs on one by one, starting with vision. The first experience would be a simple recognition of subtle light entering experience. This brain does not have a concept of memory or language, so there is no sense of self trying to interpret the light in any way. No "orange" or "yellow" or "bright"—just pure experience. Then we can bring in sounds and smells, and again with no previous experience, the brain simply allows these new sensations to enter awareness without assigning them any label. The same concept extends to the addition of touch.

All of this is an interesting way to think about an idea that has been a cornerstone of the meditative practice for thousands of years—consciousness does not need to be accompanied by thought. Indeed, most forms of meditation have the goal of being able to pull back the veil of thoughts that constantly cloud our perception of reality. Nestled within these thoughts are the labels that we place on the objects around us. Of course, these labels are necessary to be able to function in the world, but they absolutely fog the glass and I think that they take away a great deal of an objects value. The simple example that comes to my mind is the word *tree*. Because of this label, we look a forest and we see *trees*. How dare we try to explain such magnificent phenomena with single words. I try to imagine how much richer my life would be if I didn't think "tree" when I looked at one, but instead let its sighting enter my experience unhindered by previous experience.

Fortunately, this is something I can work on.

Cosmic puppets

The feeling that you are in control of your actions, that you are the conscious writer of your own thoughts, is an illusion. Everything in life—from who you marry to what you eat for lunch to your level of fulfillment at work—ultimately happens for reasons beyond your control.

The definition of free will ("...the ability to choose between different possible courses of action unimpeded") is immediately problematic. Let's begin at birth, where our possible courses of action are impeded by our genetics, our parents, and the environment we are born into. Having chosen none of these dominating aspects of our existence, how could we ultimately be in control of where we end up?

You may respond, "Sure, but the moment we become conscious we can exercise our free will by making decisions." One must go no farther than a simple thought experiment to demonstrate the lack of control we have over our decisions. Before reading on, think of a movie title. Any movie.

. . . . .

I just put you in the freest possible position you could be in. If we do have free will, certainly we are exercising it when we select a movie title. After all, it is your choice. But why did you choose the movie you did? Did you choose *Toy Story 2*? Or Miyazaki's *Spirited Away*? Why not? The point is that we don't even know why we make the choices we do—or what controls them.

Modern philosophy and neurophysiology are both more consistent with a deterministic universe than they are with a universe that provides us freedom of will. I think everything is just happening to us. We are not in control of the thoughts and sensations that enter our minds; therefore, we are not ultimately in control of the actions we take.

This moment

One of the reasons it's becoming more difficult to live in the present moment is because we have deluded ourselves into believing that life will be at its best after arriving at a set goal. An obsession with reaching an end result has certainly deprived me of enjoying the process to get there.

Viewing life as a series of stages will breed discontent. Stage 1. Stage 2. Stage 3....etc. Stage 2 will come after stage 1, and stage 3 after 2. What a wack view. With such a brief and uncertain life, I don't want to break life into predetermined segments. I'm playing a game and can't see how much time is left on the clock! Don't get me wrong, I acknowledge and appreciate the importance of routine and of obtaining the proper set of skills prior to embarking on a mission. And although certain life events do occur at similar points in time, there is no reason to think that we will be fortunate enough to experience them. The past happened and cannot be altered, and the future has not happened and is not promised. The present is the only moment that truly exists. So, what are we waiting for?

I know it is cliché, but I believe that embracing the process and enjoying the individual steps to work towards a meaningful goal is where you will find happiness. When you're attending to something you are interested in and you're engaged in it, that is when you are alive. People have recognized this basic truth for thousands of years, but as Seneca the Younger wisely stated, "Something that can never be learnt too thoroughly can never be said too often."

Climb a tree

Yesterday was solid. Woke up early, read, worked on research in a coffee shop, spent some time in the lab helping a student, and some in the office working on a manuscript and sending email. Pretty normal day for a graduate student. Despite an overall productive and satisfying time, the most joyous moment of my day came from something much simpler and much less goal oriented: the climbing of a tree.

I've been thinking about our loss of childhood wonder and curiosity as we age. I get why it happens from an evolutionary perspective, but I wish it were easier to keep that curiosity burning into adulthood. The reason is not only because curiosity drives our species into the future, but also because the feeling of curiosity is a magical one. It simply feels good to want to know how something works or where it came from. I walked past a daycare recently during recess and I saw a tiny person, probably two feet tall, sitting far separate from the rest of the children. She was sitting in the snow, running her hand through the links of the fence and seemingly counting the number of times she ran into a place where links crossed. It served as a good reminder to slow things down, be mindful of my surroundings, and to not spend my entire day in pursuit of an end goal. It inspired me (thanks tiny person).

I need to come back to my tree-climbing experience. I was jogging through Bozeman and took a detour off the sidewalk to check out some particularly nice pine trees. I checked out the bark and, upon looking up through the branches, realized it was a prime candidate for a climb. Despite this recognition, I almost didn't do it! Thoughts of the work I needed to do when I got home came into my mind and I almost left the tree standing—what a shame it would've been. I climbed the tree twice, hit my head on a branch of remarkable girth, got covered in sap, and had an overall fantastic time. It was the highlight of my day.

I hope that you can exhume some curiosity today. Climb a tree, touch a leaf, pick up a rock.

The moment is coming, Pt. I

Relatively soon, the day will come when you will get sick or someone close to you will die, and you will look back on the kinds of things that captured your attention. When we are inevitably faced with hardships like these, we will consider how we spent our time when things were normal. We may think about the times when we could have been giving our full attention to a person or group of people, but instead were distracted by something meaningless. We may think about the times we could have gone outside but did not. We may think about the time we spent occupied by petty concerns when we could have been grateful. These ruminations may not necessarily be with regret, but we will consider them.

Sam Harris discusses the paradoxical nature of this unavoidable moment in life. That is, we all know this moment of epiphany is coming, yet many of us continue to occupy ourselves with negative thoughts in this moment. It seems like a more appropriate response would be to embrace the present moment, our family, our lovers, and our surroundings to the fullest extent possible. I have always been confused as to why this is so hard. The meaning of life is happening right now and it is so obvious! But even though I can acknowledge it, write it, and say it, I will still forget it many times today and get tangled in webs of unmindful thinking. As far as I can tell, finding the true meaning of life just requires practice. It is necessary to train ourselves to pay attention to what is important—to be captured by the things that if this day were to be our last, we would be happy to have given our attention to. Because if this day is not my last, I am not far from it, and neither are you.

Today represents one precious day of the finite number that will make up my existence. I will make it count. Give my full attention to a task at hand. Feel the breath entering my lungs. Recognize the angles at which sunlight hits the snow. These moments of mindfulness will be the ones I will cherish when the end arrives.

"The end" sounds scary, but I don't think I should be afraid. To experience nothingness is something I've done before. All of us, reading this or not, share the common experience of having not existed for billions of years. It was not that scary, was it?

I think that a recognition of the fact that we are approaching the end of the game is liberating. We know that the last breath is coming. Approaching it as if it truly is the final experience of our existence can fuel a beautiful appreciation for each experience we do have. Each breath. Each sight. We are not entitled to any of them. Enjoy every experience as deeply as you possibly can—we don't know how much time is left on the clock, but we can say for sure it isn't very much.

Part II: Civilization

Collective experience

On an autumn night at a dive bar, a friend and I were discussing the shortness of life. I had recently finished reading Seneca's *Dialogues and Essays*, where he essentially claims that life is not short if you spend your time wisely. We found ourselves pondering this idea, and the conversation meandered into us considering the amount of good that an individual can do during a lifetime. It then naturally shifted towards considering what our species would be capable of if we all worked together. The following is a result of where this conversation went.

We are limited by our ~75 years of life on an individual level, but reframing to think about collective experience is powerful. If I spend 10 years on a research project, I have completed 10 years of work (minus the five years of procrastination lol). If a partner had joined me for those 10 years, we each would have contributed 10 years of our experience to the project, resulting in a cumulative 20 years of work. We each live our own life, and that makes it easy to forget that every other person on this planet lives a separate one. This means that in one calendar year, we experience ~8,000,000,000 years collectively (1 year for each earthling). Current estimates for the number of humans who have ever lived is ~108 billion. We can multiply this by the average lifespan of humans through history (~40 years) to obtain the approximate number of collective years that have been lived through history. The result is **4.32 trillion years**. Collectively, our species has lived over 300 times longer than the age of the universe. Today alone we will collectively experience over 20.6 million years of life. Imagine what is possible if we work together!

Science doesn't prove that the Earth is round

What?! Have I joined the side of the hundred flat-Earthers that have slid into my DMs over the past few years? Na. What I mean to say here is that science does not "prove" anything; the purpose of science and the scientific method is to reduce uncertainty. In the case of the flat Earth argument, science has reduced the probability of the planet being plate-like to approximately zero. Although the concept of our planet being anything but a three-dimensional spheroid is preposterous, it has become clear to me that accusing flat-Earthers of being idiots is not the correct way to address the problem. Since science works and progresses through the asking of questions and testing of hypotheses, it is important that we entertain the questions raised by this specific group of people. As scientists and thinkers in general, it is our duty to sit down and consider the arguments being made by other sides, and it is only through an assessment of bad ideas that we will be able to effectively address and hopefully correct them. In general, I see much too fast a dismissal of the flat-Earth argument—it is a bad hypothesis but an important one.

Zuck

In 2017, I had the opportunity to spend an afternoon with Mark Zuckerburg. As the founder and CEO of Facebook, he has ultimate control over what has become a cornerstone of civilization and the modern world as we know it. Whether social media is a good thing or not is controversial and there are fair points to be made for both sides. I am the first to admit that there is no shortage of problems when it comes to the effects it is having on young people especially, and I am commonly called out as a hypocrite when I emphasize the importance of getting off Instagram via Instagram (haha). Here, I just want to share something that I learned from my meeting with Zuck.

Let me start by saying that people like Zuckerberg are in no way prepared or qualified for the responsibility that comes along with creating a network of over two billion people. No one is. The CEOs who get into these positions get there through a combination of work and luck, and they could never forecast the great power and accompanying responsibility that awaits them. Zuck is a smart person, but he went from being young dude coding in a dorm room to one of the most powerful people on Earth in a matter of years. I felt in my conversations with him that he was just another computer-savvy homie who had a good idea manifest into something greater than the sum of its parts. He certainly wasn't an easy person to talk to, and that brings me to my next point. It is almost always people like Zuckerburg that are managing our social-media platforms—people who excel on a laptop but are socially inept. Isn't that annoying, that the people controlling our social media have worse social abilities than the people using it? I think that is one of reasons we should be careful with it—we shouldn't let groups of hunched over programmers tell us what it means to be socially competent. Those are only a few thoughts, but ones I think are worth pondering. It is critically important that we learn to use our new tools in a way that is societally constructive.

My experience with Zuckerberg was positive. He is a nice guy who seems genuine in his attempt to bring people together. I told him that I was into astronomy and he said, "Oh, so Elon Musk must be your hero." I responded by saying that I was impressed by the work SpaceX was doing and he told me he was still salty with Elon after a dozen Facebook satellites exploded during a failed Falcon 9 launch. Problems of a billionaire.

The scientific method

1,000-page textbooks with size-7 font and professors who got tenure in the 1980s can make it easy for students to think of science as a collection of facts. This is a sad misunderstanding, because science is not a collection of facts; it is a tool. Perhaps it would be helpful if we did not apply an overarching label of "science" to what we do, and rather referred to it broadly as "the scientific method." Thinking of science in terms of its inner mechanisms makes it easier to separate what it is from what it finds. Let's consider these mechanisms...

The scientific method consists of six parts: (1) what is your problem? (2) what do you think is causing the problem? (hypotheses), (3) how can you tests these potential causes? (experiments), (4) what do you think will happen during these experiments? (5) what did happen during these experiments? (results), and (6) what do your results tell you about your initial problem and hypotheses? (conclusions). It is the job of the scientist to follow these steps while recording information from each step and remaining aware of things that could spoil the entire process. Robert Pirsig puts it beautifully: "One must be extremely careful and rigidly logical when dealing with Nature: one logical slip and an entire scientific edifice comes tumbling down. One false deduction...and you can get hung up indefinitely." The fact that this series of steps is fragile and prone to collapse is part of what makes it so powerful. It is self-correcting, because when one or many of the outlined steps implodes on you, it is a clear indication that something needs to be reformulated.

What do the six categories above provide us in the end? Allow me to quote from Pirsig's *Zen and the Art of Motorcycle Maintenance* again, as he states, "The real purpose of the scientific method is to make sure that Nature hasn't misled you into thinking you know something you don't actually know." Science is our most powerful tool because it is applicable to everything and indifferent to our desires and biases. It will continue to tell us about Nature—how it came to be, how it operates, and how it will evolve.

Ambiguity in science

"For every complex problem there is a solution that is clear, simple, and wrong." -H.L. Mencken, 1920.

The scientific method is the best tool humans have ever created, but it does not come without sticking points. One such point exists in the ambiguities that arise naturally from the method due to the complexity of the natural world and the biases/prevailing ideas of investigators. Regarding the complexity of the natural world leading to scientific ambiguity, we can look at an example from the geosciences. In a discussion with geologist Dr. Paul Kapp, he posed the geologic questions of whether the Tibetan Plateau was shortening or extending, whether it was a contractional or extensional plateau, and whether it was rising or falling. Ambiguity surfaces when we realize that you could go either direction on all three of these questions. He went on to emphasize the importance of "...recognizing and exploring inherent ambiguities in geologic data."

Grinnell (1996) considers the role that personal biases play in scientific ambiguity, stating that, "Limited by time and money, investigators know that they will have few chances to make major discoveries...[scientists] are prepared to fight for what they believe." In other words, if ten years of your work has left you with a preferred hypothesis (regardless of discipline), it is instinctual to protect it and perhaps even disregard data that suggests something else. We can also contemplate how the structure of the research paper may lead to ambiguity (e.g., Jacob, 1988). Jacob points out the fact that papers "...replace the real order of events and discoveries by what appears as the logical order..." and are inherently flawed as a result.

For me, the wonderful (and hilarious) quote from Mencken gets to the core of some of the problems outlined here. It reminds me that nature is far too complex to be explained by a simple end-member hypothesis. That the scientific method can provide insight into the natural world, but probably never gets us quite to the Truth.

Energetic offering

Our Sun is basically a gigantic ball of hydrogen. Inside the Sun, the temperatures are so high that hydrogen atoms are heated to a plasma state, where electrons no longer orbit the protons in the nuclei. The "freed" nuclei then fuse in the process known as *nuclear fusion* to form helium atoms. In this process, not all the mass is conserved, and the small amount of mass lost is converted to energy. Einstein's famous equation, $E=mc^2$, tells us that an extremely small mass can release a titanic amount of energy (the speed of light squared is huge). Due to the massive size of the sun and incredible amounts of nuclear fusion taking place, it creates A LOT of energy. In the Sun's core, approximately $3.7×10^{38}$ protons are converted into helium nuclei every second, with around 0.7% of each fused mass released as pure energy. The resulting energy produced is ~400 yottawatts ($4.0×10^{26}$ Watts), equivalent to the detonation of ~$10.0×10^{10}$ megatons of TNT per second. Fortunately for us, planet Earth is situated at comfortable distance of 150 million kilometers from this release of energy. We therefore only receive about 1,000 Watts per square meter on Earth's surface, which is more than enough to meet our energy needs once we figure out a sustainable way to harvest it. At 1,000 Watts per square meter, it would take ~6 hours for a small backyard solar installation to provide 10 times the energy that the average American household uses in one day (West, 2017). The universe seems to be endlessly offering us entirely clean, renewable energy (wind, geothermal, solar, etc.), and we have simply not developed ways to harness and store it efficiently. Although fossil fuel extraction and greenhouse gas emissions are inevitable and necessary over the next century or so, it will be impossible to rely on them forever. Fossil fuels will become obsolete once we can reliably use renewable resources, and our harnessing of these clean resources will propel our species into the future.

Fossil fuels: a failed experiment?

A few minutes before writing this, I was stuck in traffic behind an ancient cement truck that appeared to be on the edge of simply falling apart on the spot (the type of vehicle that I like to refer to as a shi*box). The traffic started moving and the truck backfired and emitted a massive cloud of thick black smoke and exhaust from its tail pipe. The smoke that didn't end up directly in the cab of my vehicle I watched rise into the crystal-clear blue sky (haha). All by myself, I started laughing out loud—which probably doesn't seem like an appropriate response to such an occurrence. I'll explain.

Amidst the chaos of morning traffic and vehicles (like the truck above) that are constantly emitting toxic gases into our atmosphere, I was thinking about how ridiculous our current modes of transportation are—especially cars. Here's a simplified outline of how we power the internal combustion engine: Homo sapiens fight wars to gain access to the decomposed remains of plants and animals in the subsurface of our planet. We (sometimes carelessly) extract these remains (crude oil) from below and then spend huge amounts of money to refine the fossil fuels into various usable forms (like gasoline). After refining (which has a huge environmental impact), we spend billions of dollars to transport the fuel to "stations", from which WE are finally able to fill our own vehicles (at great cost to us). Over the following week, each of us drains our fuel tank and contributes to the ever-rising atmospheric CO_2 concentrations that drive global warming. The cycle continues. We don't have a solution to this complexity yet, but I think we will.

The reason I was laughing this morning is because I tried to view the truck as one would from the future. A huge metal death trap shaking and blowing noxious fumes into its surroundings. Hopefully we will look back on modern vehicles like we currently do on the horse and buggy setup.

Non-overlapping magisteria

"Science tries to document the factual character of the natural world, and to develop theories that coordinate and explain these facts. Religion, on the other hand, operates in the equally important, but utterly different realm of human purposes, meaning, and values— subjects that the factual domain of science might illuminate, but can never resolve."

In Stephen Jay Gould's *Rocks of Ages*, he first proposed the now popular idea of *non-overlapping magisteria*. Gould believed that science and religion could exist side by side, as separate but equally important tools for confronting the mysteries of natural world and human existence. Although Gould is one of my biggest inspirations and one of the best writers of popular science, his statement that science will never be able to resolve questions of human meaning and value is arguable. I like Sam Harris' approach to this in his book *The Moral Landscape*, where he proposes that the search for meaning and happiness in human life can be determined using the scientific method. His argument is essentially that once we have the technological capability, we will be able to determine what exactly would maximize meaning in our lives. He points out that simple recognition of the fact that there is a spectrum of suffering and happiness supports the idea that there must be certain actions that will maximize human well-being and ultimately meaning.

Additionally, I think Gould fails to mention important examples of biblical claims that overlap with scientific ones. The virgin birth, for example, is a religious claim—a cornerstone of Christianity. But it is also a biological claim—importantly, it is a biological claim that has no supporting evidence. On the contrary, the idea of a virgin birth contradicts everything we have ever observed in the natural world. This is one example where biblical truths contradict scientific fact and there is a breaking down of non-overlapping magisteria.

The problem of evil

There are three explanations for evil in the world. The first is that god is not all-powerful and cannot control how the universe works, the second is that god is evil and creates misery and pain for its own sake, and the third is that god does not exist. This line of reasoning was popularized by the philosopher David Hume, who presented the following questions:

Is God willing to prevent evil, but not able? Then he is not omnipotent. Is he able, but not willing? Then he is malevolent. Is he both able and willing? Then from whence comes evil?

The problem of evil, which questions how we can reconcile the existence of suffering with an all-powerful, all-good, and all-knowing supreme being, has stood strong at the center of philosophy for thousands of years. Arguments have ricocheted from it at all angles, from those that deny the existence of evil to those to those that claim evil disproves god's existence. One common argument is that evil arises because of human free will—something I don't subscribe to considering that free will seems to be illusory. Another (perhaps more far-fetched) argument is that of *pandeism*, or the theory that asserts god created the universe and subsequently became the universe. Many similar lines of reasoning propose that god no longer exists "above" and does not have the power to intervene and prevent and/or reduce human suffering.

There is a lot of uncertainty surrounding how to explain evil, but one thing seems certain—evil exists. I am unconvinced by "privation theories of evil", which assert that evils such as suffering and disease only appear to be real, but in truth are illusions, and in reality evil does not exist. One has to look no further than the point made by the writer Stephen Fry that there are insects on Earth whose entire life consists of burrowing into the eyes of children to make them blind. I suppose the bottom line is that if you believe that god does exist, you have to ask yourself, "What kind of god is he?"

Things are getting better

It seems like the consensus around the world is that things are getting worse. This interpretation is a result of the fact that our information about what is wrong with the world is getting exponentially better. Just a few centuries ago, it would have taken weeks or months for us to learn of the destruction of a village separated from our own. The technological advances that have occurred over the past fifty years have enabled us to learn instantaneously what is happening across the world. Today, we are informed within seconds when anything bad happens anywhere on Earth. Think of how difficult that abrupt change must be on our slowly evolved human brains. Paired with the mainstream media's business model of dramatically reporting news that is designed to evoke feelings of fear, this exploding technology misinforms us on the current state of reality.

Not only are things the best they have ever been in human history, but they are constantly getting better. Statistics across the board display dramatic positive increases that represent empirical evidence of our progress. Over the last 25 years, there has been a 69% decrease in extreme poverty in developing countries across the world (World Bank). In 1915, 10% of the world population could read and write. Today, the literacy rate approaches 90% (World Bank). Meanwhile, global life expectancy has increased from 31 years (!) to almost 80 (World Bank). Things are getting better, we are simply not informed by our media outlets. Fortunately, there are organizations that recognize this truth and allow us access to these reassuring statistics (humanprogress.org).

I understand that there are still huge problems, massive injustice, and horrific violence across the globe that need to be addressed and fixed. However, we can still be grateful for the improvements we have made to this point, they are truly amazing.

No control

In Part I, we discussed the illusion of free will. Put simply, we have no control over the external events that cause us to take certain actions (nor do we even have control over the thoughts that arise in our heads). Now, I want to consider the illusion of free will in the context of political disagreement (and differences of opinion generally).

In the United States, we've spent lots of time in a contentious political situation. Growing polarization of society between two political parties (neither of which is that respectable) has created a country with an ideological split down the middle. Most political discourse in this country consists of each side spewing insults at the other and stating that their answers to our problems are so obviously the right ones! At least that is what the news does. Bottom line, it is filled with contempt and is unpleasant. I am kinda exhausted by people who despise others for their political beliefs. I think that recognizing free will as an illusion makes the anger and frustration towards people with opinions that differ from your own become subdued. You realize it is ridiculous to be angry at a person for something that they have no control over.

Think of a person that you cannot stand because of their political stance, someone who you think has their values all backwards. Now imagine that you had their parents, their education, their life experiences, and their neurophysiology (all things that they have no control over). You are now them, and you have their opinions. Doesn't this make you think that your anger may be misplaced?

This is not to say that you shouldn't fight for what you think it right. You absolutely should. But no problems can be effectively solved when the problem solver is driven by resentment and anger. I think it is more effective to approach these kinds of things with compassion. Seneca the Younger once saying in a letter to his friend Lucius, "No one gives advice at the top of his voice."

Certain political actions are inherently angering, particularly ones that hurt people. I'm not saying you shouldn't be angry—but I'm saying that you should recognize what it is rooted in. And we shouldn't let it be the driving force behind our actions. Whether or not you think that this is a good approach to politics is also out of your control.

Part III: Earth

Wegener's missing mechanism

On September 7, 1963, a paper was published that would transform our understanding of Earth and its mechanisms. This is a plate tectonic story—one that confronts, among other things, how mountains are formed.

The story begins with centuries of misled geodynamic thought, beginning with a static view of the planet in which the Earth and its continents appear exactly as they did when God created them. The earliest challenge of this view came from "mobilists"—people who recognized the congruency of coastlines (Africa and South America, for example). Sir Francis Bacon was one of the first to suggest a former union of continents, recognizing that they, "...fitted together and must be related." (Bacon, 1620). Numerous early scientists acknowledged the congruency in the intervening 300 years, but it wasn't until 1912 that an actual theory was put forward to explain it. This theory—Alfred Wegner's theory of continental drift—suggested that Earth's landmasses had once been conjoined and had since fragmented. Like many revolutionary scientific ideas, Wegener's was met with skepticism. This was largely because (despite all the evidence he had) he lacked a viable explanation for the driving force of continental drift. Fast-forward a few more decades, and we come to the year 1963, and the publication of *Magnetic anomalies over oceanic ridges* by Fred Vine and Drummond Matthews.

This paper, emerging out of advances in marine geology/geophysics, paleomagnetism, and seismology, suggested the following: the ocean floor is imprinted with the record of magnetic field reversals in the form of alternating stripes, with widths proportional to intervals of the polarity reversal scale, formed by seafloor spreading from mid-ocean ridges.

There is a lot there, I know. But if you reread it a few times its beauty may become apparent. It simply proposes that reversals in Earth's magnetic field are recorded in the rocks of the spreading sea floor. Broadly speaking, the hypothesis provides the driving force that Wegener was always missing (that is, sea floor spreading). I truly believe that the Vine-Matthews-Morley* hypothesis represents one of the greatest displays of human thought, with an elegance and importance comparable to the great theories like Einstein's relativity or Darwin's evolution by natural selection.

Mighty mountains

We are fortunate to live on the only planet in our solar system known to have active plate tectonics. The lithosphere of Earth is broken into several different "plates", each of which can move because of their higher mechanical strength relative to the asthenosphere and mantle beneath them. As plates of oceanic and continental lithosphere interact, mountains are built. The convergence of a dense oceanic plate with a less dense continental plate commonly leads to subduction, in which the oceanic plate is thrust underneath the continent and is eventually gobbled up by the mantle. An example of this type of convergence would be the west coast of South America, where the Nazca Plate is subducting beneath South America and creating the Andes. The big, beautiful chain of volcanoes that trace the axis of the Andes is also a result of this subduction.

The collision of two continents also creates enormous mountains (e.g., the Himalayan-Tibet orogen). Since India's collision with southern Asia approximately 58 million years ago (Kapp & DeCelles, 2019), the Greater Himalaya have grown to

---

* Although Vine and Matthews published the first paper suggesting this hypothesis, another geologist (Lawrence Morley) came to the same conclusion independently. In fact, he was the first person to send the idea into peer-review in early 1963 (both submissions were rejected).

become the highest mountain range on Earth. 60 million years has not been adequate time to stop the northward migration of the Indian subcontinent. As a result, the Himalaya are an example of an active collisional mountain belt (one that is still rising). Also a result of this collision (and the preceding closure of ancient ocean basins and accretion of terranes) is the Tibetan Plateau, which with an average elevation of ~5,000 meters is the highest and largest plateau on Earth. This "roof of the world" is arguably the most interesting area geologically on our planet and it is likely that its uplift led to changes in global climate.

The study of mountains is immensely interesting and fulfilling. We are a lucky generation—the ones who are fortunate enough to live at this time on this planet with the problems of its mountains.

<u>Mighty mountains Pt. 2</u>

I find few things more awe-inspiring and humbling than the mountain ranges on Earth. Usually formed over tens to hundreds of millions of years, mountains are created by the unparalleled power of plate tectonics. As plates scrape past one another, diverge, and crash together, the rocky crust of our planet is torn, exhumed, deformed, uplifted, and eroded. Mountains like the great Himalaya rise 10% of the way to the Earth-space boundary, and multi-billion-year-old ranges that were once of equivalent size have been eroded and their metamorphic cores are now exposed at sea level. The rock cycle is untiring and the mountains we see today are a miniscule snapshot in geologic time—a fraction of a frame in the great movie of Earth history.

I find inspiration writing this in the shadow of the Bridger Mountain Range in southwest Montana. Being familiar with the geologic history of my surrounding area, I know that the range formed through a series of tectonic events between the Late Cretaceous and early Eocene (between 80 and 55 million years ago). Over my morning coffee, I'm considering the history of these mountains on a human time scale and contemplating that they probably look nearly identical to how they did when our Mesopotamian ancestors were creating the first written language and inventing the wheel. Sure, maybe slight differences in drainage geometry and vegetation, but the structural framework of the mountain range has not changed significantly since the foundation of the Sumerian civilization 8,000-10,000 years ago. Similar to what the great Carl Sagan once said, think of how much blood has been shed, how many leaders have risen and fallen, and how many empires have been established and dissolved since the first human civilization. The mountains remain as they were, unphased by our petty arguments, overly complex human brains, and insatiable hunger for change. It is for this reason, among others, that I am constantly humbled by the mountain, the dominating feature of Earth's land.

Why do rivers form?

To approach the question above quantitatively, let's simplify a river to what it is in essence—a flow with unchanging discharge and uniform depth contained in a rigid channel with uniform slope (e.g., Allen, 1985). This setup is shown schematically below. We are now prepared to consider the driving and resisting forces of a given flow in the context of

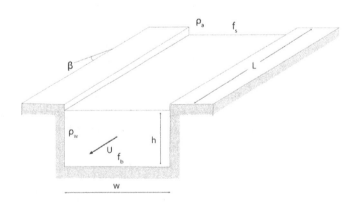

Newton's first law of motion (assuming uniform and steady flow, driving and resisting forces must balance; Eq. 1 & 2). The driving force of our flow is a function of the flow volume, the flow density, gravity, and the slope of the river. The resisting force is primarily frictional drag that the flow experiences along its "wetted perimeter" (where the flow is in contact with the channel walls). Importantly, our resisting force is mostly dependent on basal shear stress (τb), something that will come back to inform our question about why channels form. Setting these equations equal (Eq. 3) and doing simple algebra allows us to solve for shear stress.

A few notes on the simplification:

- L (channel length) cancels right away.
- Basal shear stress is substantially larger than the shear stress of air, therefore we disregard τa.
- We are assuming a river much wider than it is deep; therefore, the first part of our expression for τ simplifies to h (studying Eq. 4 will help understand).

$$F_{driving} = (whL)(\rho w - \rho a)g\sin\beta \qquad (1)$$
$$F_{resist} = (wh\tau a) + L(2h + w)\tau b \qquad (2)$$

since $\tau a \gg \tau b$, disregard τa

<u>Setting Fd=Fr</u>

$$(wh\rho w)g\sin\beta = (2h + w)\tau b \qquad (3)$$

$$\tau b = \frac{wh}{2h + w}\rho wg\sin\beta$$

since $w \gg h$, $h\left(\frac{w}{2h+w}\right) = h(1) = h \qquad (4)$

at small β, $\sin\beta = \tan\beta = $ slope (S)

$$\therefore \ \boldsymbol{\tau = pghS} \qquad (5)$$

- Finally, the angle of our flow below the horizontal, represented by $\sin(\beta)$, is equal to $\tan(\beta)$ at very low angles. And $\tan(\beta)$=slope of channel. Therefore, $\sin(\beta)$=slope.

Things are much nicer now. In words—shear stress is equal to the product of flow density, flow depth, flow slope, and a gravitational constant (Eq. 5). This is profound, and it answers (with caveats) our original question. It tells us the following: a small imperfection on an otherwise flat plane locally increases depth (h) and slope (S), which consequently increases shear stress. Increased shear stress leads to erosion, accelerated feedback begins, and a channel is born.

Order within chaos

Nature is full of apparently random disorder and irregularity; it is chaotic. This term *chaos* can be misleading however, because it carries with it a connotation of complete randomness and disorder. On the contrary, chaotic behavior hides order; it is simply highly dependent on initial conditions and is deterministic. If tiny changes are made to the initial condition of a system, the outcomes will diverge widely. Let this bring us to consider the chaotic behavior of a meandering river system.

Stølum (1996) wrote an excellent and thought-provoking paper titled, *River Meandering as a Self-Organization Process*. In this work, he runs simulations of freely meandering rivers and demonstrates that the process of meandering self-organizes a river into a state characterized by fractal geometries. He observed oscillations in the system between ordered and chaotic states that was dependent on channel cutoffs (cutoff of a meander; formation of oxbow lakes). When cutoffs occur in an ordered state (strong axial symmetry), they tend to change the system into a chaotic state (low symmetry). This transition from ordered to chaotic is show here in comparison to smoke rising from the end of a lit cigarette. Technicalities aside, the study serves as a good example of how natural processes create their own order.

Consider the sensitive dependence on initial conditions for both processes shown on the following page. Why does the transition from order to disorder occur when it does? After becoming chaotic, what processes are behind the turbulent dispersal of material? Without getting into complex mathematics, we can say one thing for sure: these systems would not look the same if we made any alteration to their initial conditions.

If I may...everything in the cosmos is just happening. It has unfolded and will continue to unfold in the one and only way possible given its initial conditions. Nature has a fractal pattern—there is no reason to think that the dependence on initial conditions stops at the scale of a river or a mountain

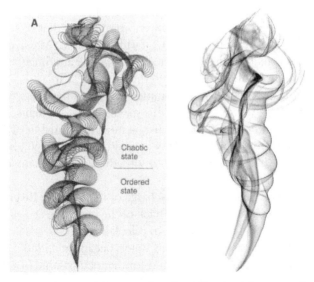

Comparing modeling results of a self-organizing meandering river system (left; Stølum, 1996) and a wisp of rising smoke (right).

range. Perhaps consciousness in the universe is semi-analogous to how a river cutoff can change a system from chaotic to ordered...regardless, we are not separate from nature in any way.

The weight of water

Water on water on water (s/o Young Dolph). Of the many things that we take for granted, it is my humble opinion that none are as underappreciated as water. It goes without saying that the existence of carbon-based life hinges on the ubiquitous presence of H_2O. However, I have been thinking more lately about the sheer power, force, and mass of Earth's oceans, lakes, and waterways.

I once had the privilege to drive along the Yarlung-Tsangpo River. This river forms the southern boundary of the Tibetan Plateau and becomes the Brahmaputra, ultimately flowing into the Bay of Bengal. With an average discharge of over 500,000 cubic ft/sec, it is a truly enormous river that is incising itself into the southern plateau with authority. Massive lateral waves project themselves across the thalweg and huge hydraulics boil out of the depths and then disappear in a suction-like fashion. Constriction rapids curl and smash the water upstream. As I drove along this spectacle, my thoughts landed on how much pure mass must be moving through it. I considered how interesting and easy it would be to calculate the mass of the discharge, then scale it to something more relatable than just a number.

I did this, and the mass discharge of the Yarlung River was indeed impressive. But many back-of-the-envelope calculations ultimately brought me to calculating the approximate total mass of all Earth's water, oceans included. The most consistent estimate I found for the total volume of water on Earth was 333 million cubic miles (please excuse the incoming abuse of the imperial system). I began by determining how many gallons were in a cubic mile, simply because I am familiar with the mass of one gallon. 1 cubic mile is equal to 1.1

trillion gallons, and at 8.34lbs/gal, has a mass of ~9.2 trillion pounds. Multiplying this value by the estimated 333 million cubic miles of water on Earth gives one a value 3.0636×10^{21}. That is ~3.1 sextillion pounds (1.4 sextillion kilograms) of water on Earth! Fantastic!

As I said, I wanted to make such a huge number more relatable, so I decided to calculate the total mass of all human beings on Earth. Taking the average human mass of 137 pounds and multiplying it by ~7.7 billion people, the total mass of the population is around 1.1×10^{12} lbs (4.9×10^{11} kg). Simple division then tells us that people are outweighed by water 2.8 billion times (and this does not even consider the fact that we are mostly made of water haha!).

For me, this comparison serves as a reminder of the vast size and power of nature. Earth's water will continue to amaze me both as life's elixir and a sculptor of the landscapes we inhabit.

The weight of clouds

I've always envied the clouds, floating across the sky with no destination. I look up to them as they interact with nature in a way that I long to—peacefully. Without resistance. I recently came to the realization that I've treated clouds in an almost metaphysical manner for most my life. I think it's primarily because they seem weightless, and weight is so fundamental to the way we observe and classify physical objects. But of course, like all things that may seem metaphysical, clouds are not. Clouds are finite and bound to nature by the same laws that govern the rest of the cosmos; therefore, as much as I wish they didn't, they too have mass. In fact, clouds are heavy! Let us explore.

Clouds are made of water. When the gas form of water (water vapor) rises in the atmosphere, it progressively gets cooler and is under less pressure. Decreasing the temperature and pressure of the air reduces the amount of water vapor it can hold; as a result, the vapor turns into water droplets or ice crystals and a cloud is born.

Not all clouds are created equal. They come with varying density that is a function of the amount of water they hold. For the purposes of our thought experiment, let's consider the classic bubbly cumulus cloud, which has an average density of ~0.5 grams/cubic meter. Knowing the density, all we need is an approximation of the size to determine the clouds mass. Assuming a close to spherical cloud that is 1,000 meters long, wide, and tall gives us a whopping volume of 1,000,000,000 cubic meters (not an uncommon size by any means). Using the density equation, we get a mass of 500,000,000 grams (500,000kg; 1,100,000lbs). Over a million pounds! Of course, the mass of a cloud scales with its size! An average cumulonimbus (thunder cloud) has a mass easily into the millions of tons. Magically, all that mass is pushed through the Earth system as part of the water cycle that we owe our lives to.

Even though clouds are clearly bounded physical entities, there is still something about them that can't be explained entirely by physics. That is, the feeling I experience when I look at them. Feelings of calmness. To me, they reflect my own consciousness back to me. Perhaps if all matter is imbued with consciousness, as some scientists believe, they are conscious themselves—looking down on us as we look down on ants.

The weight of mountains

As my buddy and I looked out over a spectacular and rugged landscape in southwest Montana, we pondered the weight of mountains. We considered the difficulty of picking up a big rock (simpletons lol) and imagined how many times its mass would have to be multiplied to approximate even a single peak within a mountain range. How much do mountains weigh? Back of the napkin calculation coming in hot.

The easiest method to approximate the weight of a single mountain is by creating an idealized model, like we did for our river a few entries ago. For this

$$V_{mtn} = \pi r^2 \frac{h}{3} \qquad (1)$$
$$v = \pi h^3 \qquad (2)$$

∴ **for our experiment**:
$$v = \pi(3000m)^3 = 8.4 \times 10^{10} \ m^3$$

calculation, we will assume a perfect conical shape for our mountain and a single rock type (i.e., homogeneous rock density). Our idealized mountain is made entirely of granite, has a height of 3,000 m, and has sides with a consistent steepness of 30-degrees. The volume of a cone (Eq. 1) can be simplified for our 30-degree mountain to Eq. 2 (see footnote). With a height of 3000 m, we calculate a volume of 8.4×10^{10} cubic meters. At 2700 kilograms/cubic meter, we can then do simple multiplication to determine a mass for the idealized mountain of 2.29×10^{14} kilograms. A large number indeed, but it doesn't mean much unless we scale it to the mass of something we're familiar with.

For perspective, let us now compare the mass of a single mountain to the mass of all living humans on Earth. The average mass of a human is around 62 kilograms and there are 7.7 billion people on Earth—this allows us to calculate the total mass of humanity at $^{\sim}4.8 \times 10^{11}$ kilograms. Simple division can then tell us that the total mass of humanity is 0.208% of the mass of a single mountain. Worth repeating: *the total mass of humanity is 0.208% the mass of a single mountain.*

We are deluded in thinking that we are in any way, shape, or form the dominant force on this planet. We are just creatures, dwarfed in every way by the planetary and cosmic

---

* If a mountain has height h, it's radius is 1.73h or h$\sqrt{3}$. Therefore, if we consider the mountain to be a cone of height h and radius h$\sqrt{3}$, then the volume is $\frac{1}{3}(base)(height)$ or $\frac{1}{3}\pi 3h^2 h$. This simplifies to Eq. 2 ($v = \pi h^3$).

phenomena that surround us. Let's fill our role in a humbler fashion.

The rock cycle

The planet we inhabit is a restless one. Constantly changing crustal stress regimes due to plate tectonics, a fluctuating climate, and physical and chemical weathering at the surface are a few of the dominating artists that contribute to the sculpture that is planet Earth. As streams and rivers attempt to strip the land of its crust, the mighty volcano offsets their work by contributing new crust to the system.

Along with producing some of the largest topographic features (e.g., Cerro Aconcagua in Argentina), volcanism can produce some of the most incredible sights in the solar system, especially on our home planet. With active plate tectonics and an abundance of internal heat, planet Earth has lava, pyroclastics, and volcanic gases constantly escaping through vents at its surface. Our satellites and spacecraft in low-Earth orbit give us incredible perspective on the clouds of gas and ash that rise from these openings in the crust. Once material makes its journey from the depths of our planet to the surface, it immediately becomes vulnerable to the atmosphere. Ultimately these volcanic rocks are broken down, transported, deposited, re-solidified, and possibly even re-erupted from a new volcano millions or billions of years later—all part of the beautiful, never-ending rock cycle.

I came across a study that represents a beautiful demonstration of the rock cycle as described above. This study (Jacobson et al., 2011) was broadly interested in the tectonic evolution of southern California. The researchers used small detrital minerals called zircons to investigate the provenance (source history) of sediments that were deposited in a trench setting offshore of modern-day California. These detrital zircons are tough little bastards who can survive at high temperatures and pressures within the Earth. Their study found that zircons in the trench strata were shed from an exposed granitic rock at the surface, deposited on an oceanic plate to the west, subducted (or pushed) underneath the North American continent, and then re-exhumed to shallow crustal levels, all within ten million years! A blink of the eye on a geologic time scale! Earth rules.

The golden spike

Human activity has had and continues to have a significant impact on the environment, so much so that the numbers describing it have become kinda scary. In a paper titled *Defining the Anthropocene* (Lewis and Maslin, 2015), a handful of examples are given to emphasize this point. The authors point out that since the invention of the process that converts atmospheric nitrogen to ammonia for use as fertilizer (the "Haber-Bosch process"), we have altered the global nitrogen cycle in a way more dramatic than any geologic process has since 2.5 billion years ago. Our appropriation of ~30% of net primary productivity for human use has resulted in species extinctions between 100 and 1,000 times the background rate (perhaps the beginning of the sixth major extinction event on Earth). Furthermore, we have released approximately 555 billion metric tons of carbon into the atmosphere since 1750, causing continually increasing average global temperatures, higher ocean water acidity, and a likely delay in Earth's next glaciation event. Science makes it clear that we are changing Earths biogeochemical cycles and the evolution of life.

One of the reasons that Lewis and Maslin (2015) describe these numbers is to contextualize the place of our species on a geologic time scale. More specifically, they outline how these types of actions may justify the classification of a new "...human-dominated geological epoch, the Anthropocene." The Anthropocene is a term that is now commonly thrown around by science writers and everyday people, but there is not actually an agreed upon start date for the new epoch (dates ranging from before the end of the last glaciation to the 1960s). In fact, some do not believe that the human impacts on the stratigraphic record are prominent enough for there to be an Anthropocene at all.

Formal geological time units are defined by their lower boundary and are demarcated using a Global Stratotype Section and Point (GSSP). These sections and points are locations within the geologic record (rock, sediment, ice, etc.) that contain physical or chemical changes reflecting a global change phenomenon. The well-known K-T boundary, broadly representing the extinction of dinosaurs, has its GSSP located in ~66-million-year-old rock in El Kef, Tunisia (for example). According to the geologic time scale (Gradstein et al., 2012), GSSPs (also known as *Golden Spikes*) must have the following characteristics: (1) a principal correlation event (the marker), (2) other secondary markers (auxiliary stratotypes), (3) demonstrated regional and global correlation, (4) complete continuous sedimentation with adequate thickness above and below the marker, (5) an exact location—latitude, longitude and height/ depth—because a GSSP can be located at only one place on Earth, and (6) be accessible.

To define the Anthropocene as a formal geologic unit of time, it must be marked in the geologic record (within rock, sediment, ice core, etc.). Where (if anywhere) to place the Golden Spike for the Anthropocene remains a topic of discussion. Obviously, before marking the GSSP, it is necessary for scientists to identify a date that conform to the criteria above. According to Lewis and Maslin (2015), there are two such dates: 1610 and 1964.

The 1610 possibility represents the "collision of the new- and old-worlds", when Europeans arrived in the Americas. Trade networks linking Europe, Asia, and the Americas led to the globalization of food trade and a decrease in atmospheric CO_2 due to a massive decrease in people (from 60 to 6 million; disease, famine, war). This decrease in CO_2 between 1570 and 1620 is preserved nicely in Antarctic ice cores—a plausible GSSP.

The year 1964 represents an unprecedented increase in human population, creation of novel inorganic materials such as plastic, and the tail end of decades of nuclear bomb detonation and its associated radioactive fallout. The massive amount of chemicals put into the Earth system provide another possible starting point for the Anthropocene. These gases (most importantly, carbon-14) are captured by tree rings, leading Lewis and Maslin (2015) to suggest a specific pine tree outside of Krakow, Poland as the 1964 Golden Spike. As of May 2023, the Anthropocene still lacks a formal geological definition (Voosen, 2022; Schwägerl, 2023).

Respect the drip

The Earth's crust is dripping! What the hell does that even mean?

We are all familiar with the magnificent structures created by plate tectonics. The mighty mountain belts of Earth rise dramatically from valley floors into the low levels of the atmosphere, representing the culmination of tens to hundreds of millions of years of contractional and extensional stresses. Although mobile tectonic plates are mostly responsible for crustal deformation and Earth's largest features, there are also deep lithospheric processes that can express themselves at the surface. One particularly interesting geological phenomenon filling this description is known as lithospheric drip or *drip tectonics.* This phenomenon can generally be thought of as detaching and foundering pieces of the lower lithosphere into the greater depths of the planet. Think of them as falling tear drops.

To begin to understand drip tectonics, one must first grasp the structure of the Earth's crust. We commonly think of the outermost shell of Earth as the crust, which is mobile atop a semi-molten mantle. It is a little more complex than this. *Lithosphere* is the term used for the outermost rigid crust and the elastically behaving upper-most portion of mantle (known as the lithospheric mantle). In drip tectonics, it is understood that the lower lithosphere can become negatively buoyant relative to the underlying upper mantle (e.g., Molnar & Houseman, 2004). As a result of this density contrast, the mass of lithosphere begins to sink into the mantle and, over a period of ~10 million years, detaches and founders into the depths. This removal of lithosphere can express itself at Earth's surface in the form of subsidence (a gradual sinking of an area of land). The drips literally pull down on the surface of Earth and can create space for sediment to accumulate (e.g., DeCelles et al., 2015; McMillan & Schoenbohm, 2023). In addition, after a drip detaches there is a rapid increase in surface elevation as the crust isostatically adjusts itself. I am a geology nerd, but I think that is one of the most intriguing displays in all of nature.

Praise be to the mighty Z

The zircon crystal is a microscopic mineral that is commonly used in the geosciences to determine the age of a rock. Zircon ($ZrSiO_4$) has an atomic structure that, during its formation, gladly welcomes the radioactive isotope uranium (U) into it, while refusing to accept the relatively large lead (Pb) atom. After crystallizing in a magma chamber, the radioactive uranium atoms are locked into the crystalline lattice of zircon, and over the course of millions or billions of years, radioactive decay of uranium to its daughter product (most commonly $^{238}U \rightarrow ^{206}Pb$) leads to the accumulation of lead in the zircon crystal. Knowing the decay constant/half-life of uranium, it is possible to measure the relative abundances of parent/daughter to obtain a U-Pb geochronologic age.

Over the past few decades, technologic advances and a better understanding of the U-Pb system have allowed for the development of this technique to an unprecedented level. The most common and efficient way that U-Pb ages are measured from zircons is through laser ablation inductively-coupled plasma mass spectrometry (LA-ICPMS). In this technique, individual polished zircon crystals are shot with a laser (~30 µm beam diameter), the ablated material is ionized by sending it into a plasma torch, the ions are sent through a mass spectrometer (essentially a magnet), which separates ions (U & Pb) by atomic mass, which are finally measured by a series of detectors on the other side. With a few corrections for various errors, the ratio of U/Pb can be converted to an age in millions of years with 1-2% uncertainty. Even though that might have sounded technical, we are essentially shooting crystals with a laser, measuring the abundance of different atoms, and obtaining an age. This system outlined above is largely responsible for our current understanding of the age of the Earth and solar system.

From gaining insight into the tectonic evolution of mountain belts to unraveling the secrets of ancient river systems that once transported sediment across Earth's surface, geochronology will remain one of the most powerful techniques we possess to investigate the Earth system.

A shared planet

From the depths of the North Pacific Ocean, a 170,000-kilogram blue whale rises through a swarm of krill on its way to break the surface. In east Africa, giraffes wander across the rolling grasslands of the savanna with no particular destination. Meanwhile, herds of wild yak graze on sedges atop the wind-swept and immensely vacant Tibetan Plateau. On the other side of the planet, a poison dart frog takes cover from deluge beneath a waxy leaf on the floor of the Amazon Jungle.

One of the greatest pleasures in my life is knowing that I share the planet with the diversity of other creatures. I try frequently to conjure images like the ones above to remind myself of what magical biological and ecological events are always taking place. Personally, seeing animals in nature and reading about their habitats and behaviors drives home the responsibility we have as an intelligent species to protect them. Our societies place such a strong emphasis on maximizing human well-being that it can be easy to forget that there are animals of absolute grandeur and importance that need to be considered as well. As an intelligent species that uses a lot of resources, it is truly our responsibility to protect and respect these creatures. It is a moral obligation.

The diversity of life

The diversity of life on Earth is remarkable. Some 10,000 species are discovered each year, with estimates for the total number of species ranging from 8.7 million to 1 trillion (Sweetlove, 2011; Locey & Lennona, 2016). This diversity has always fascinated me. If you were not raised on Earth, imagine seeing this chameleon next to a whale shark or a silverback gorilla. Wouldn't you be absolutely amazed that each of these species evolved on the same planet? It is a true testament to the power and magic of evolution by natural selection.

In Part I, we discussed how easy it is for one to not fully appreciate things whose presence is consistently there. The longer something is there, the easier it is to assume it will stay. We have evolved alongside these creatures over hundreds of millions of years and it seems that many of us take their exquisite beauty for granted as a result. I think we all need an occasional reminder that we share the planet with other species that have worked equally hard evolutionarily to get where they are today.

Also, when pondering life on other worlds we need to keep in mind the incredible range of organisms that evolution can produce. To me, the animals around us look more alien than any image I can conjure up in my mind.

All hail

Creatures, with morphologies and behaviors far stranger than anything we could conceive, have us surrounded. Scuttling, flying, excavating, diving, swimming, and breathing creatures. I love them. Many of the strangest animals seem to fill the gaps between where humans reside. They don't like us too much. I don't blame them. They need a nature unperturbed by human activities to thrive. Perhaps their oneness with nature is one of the things that make them so beautiful and so intriguing.

Imagine I'm describing a spider to a being with no prior knowledge of Earth's biodiversity: "We live with a creature that has eight long and skinny legs, four on each side. It has four forward-facing eyes in a row, the middle two of which are twice as big as the ones on the outside. Two small appendages attached to the head allow it to push prey into its fangs for feeding. It is dazzling in color—dark glowing purples, iridescent yellows, and blood reds. We call it a spider." Or perhaps a different creature: "This one has a thick, rubbery, oily skin exposed on its head and limbs, the latter of which are known as 'flippers'. The rest of its body is encased in a large, bony, durable shell that is bilaterally symmetrical. They live in the ocean but cannot breathe underwater. They overcome this challenge by holding their breath for seven hours at a time, slowing their heartrate down to one beat every nine minutes. We call it a sea turtle."

In both cases, we might as well be describing the most ridiculous fictitious alien we can think of. This same thought experiment goes a long way for most species on the planet. Planet Earth's great biodiversity makes my life richer and more worth living. All hail the mighty creature.

Bioluminescence

Aboard the *HMS Beagle*, Sir Charles Darwin wrote the following in his journal:

"While sailing in these latitudes on one very dark night, the sea presented a wonderful and most beautiful spectacle. There was a fresh breeze, and every part of the surface, which during the day is seen as foam, now glowed with a pale light. The vessel drove before her bows two billows of liquid phosphorus, and in her wake she was followed by a milky train. As far as the eye reached, the crest of every wave was bright, and the sky above the horizon, from the reflected glare of these livid flames, was not so utterly obscure, as over the rest of the heavens."

The pale glow of the sea and illuminated wave crests that Darwin describes here is one of the first field observations of the phenomenon known as bioluminescence. In its simplest form, bioluminescence is the emission of light from living organisms through chemical reactions in their bodies. This reaction requires two unique chemicals known as luciferin (substrate) and luciferase (enzyme). The former is the compound that produces light, and the latter is an enzyme—a chemical that interacts with the substrate to change the rate of chemical reaction. Generally speaking, the reaction between luciferin and an oxidized luciferase creates two byproducts: a new compound called oxyluciferin and, most importantly for today's discussion, light! The beautiful colors we see in bioluminescent organisms (e.g. blue in phytoplankton, yellow in fireflies) is controlled by the configuration of the luciferin prior to the reaction taking place.

Bioluminescence occurs across thousands of species, ranging from beetles to fungi. It is most common in marine animals, especially creatures of the deep. Light can range from small spots and patterns to an illumination of the entire creature. There are many evolutionary reasons for bioluminescence in nature, but the most common include camouflage, attraction and deterrence of others, self-defense, communication, mimicry, and illumination of the surrounding environment.

I love trees

A few hundred years before the ancient Egyptians constructed the Great Pyramid of Giza, a seed germinated in the soil of western North America. Today, almost 5,000 years later, that seed lives on as Methuselah—the oldest tree on Earth. This tree came into existence when the world population was less than 14 million people, lived through the invention of alphabetic writing, and was standing strong at over 2,000 years old during the rise and demise of the Roman Empire. I love thinking about space, time, earth, and nature for these types of humbling reminders. And through everything I've learned about geologic time, the vastness of the cosmos, and the depths of the oceans, few things have inspired me more than the mighty tree.

I love trees. Having existed as long as many of the rocks on Earth, they have a rich and complex evolutionary history and have played a key role in the development of biological life. Seemingly too good to be true, absorbing the now problematic gas of carbon dioxide from the atmosphere is what allows them to live. The photosynthetic byproduct of their survival is oxygen, a gas that is directly responsible for supporting other life on the planet. Aside from enabling us to live, many of their characteristics seem simply magical to me. The giant redwood (*Sequoiadendron giganteum*) serves as a good example, as it is capable of turning from a tiny seed weighing less than one gram to an 85-meter (275ft) tall, 1.2 million-kilogram (2.7 million lb) behemoth. A single quaking aspen tree (*Populus tremuloides*) exists in south-central Utah, USA that occupies 106 acres and has a weight of 6 million kilograms, making it the heaviest organism on the planet. A substantial body of scientific evidence now suggests that trees of the same species can **communicate and form alliances** via interconnected fungal networks that exist a few inches beneath forest floor (e.g., Simard & Durall, 2004).

Trees are truly fascinating organisms whose importance should not be valued lightly. We are fortunate to share the planet with organisms of such grandeur.

Sharing the planet

I have always been fascinated by biology and ecology, I think largely because of their complexity. Like Earth's huge mountains and gushing rivers, the three domains of life serve as humbling reminders that we share this planet with things much greater than ourselves (or at least they should). Above, I acknowledged the fact that we exist alongside other species that have been around far longer than us. In fact, our species is a relative newcomer on the stage of life. A clarifying example: our ~200,000-year existence represents 0.0363% of the amount of time that jelly fish have occupied the oceans. Even more impressive is the cyanobacterial stromatolite that is still thriving in modern waters. Existing in the same form for ~3.0 billion years, the stromatolite has been around 15,000 times longer than Homo sapiens. These kinds of examples could continue indefinitely.

Our complex brains and recent advances as a species have deluded us into thinking that we are the most impressive occupant of planet Earth. I don't think we are. From a biological perspective, longevity can serve as a proxy for fitness, and the kinds of examples above make clear that we have not yet earned any bragging rights. The recognition that we are sharing the planet with all other species is not only magical, but essential for the future protection of biodiversity.

I'm gonna go outside and pay respect to nature. I'm gonna approach a tree for what it is—something that is taller, stronger, older, more beautiful, and possibly even smarter than I am.

Animal weapons

The surface of our planet is teeming with creatures that inspire awe in their form and nature. A result of the mechanism of evolution by natural selection, many of these organisms boast oversized animal weapons that I feel are largely underappreciated as some of the most ridiculous of all natural phenomena.

Douglas Emlen's book *Animal Weapons* consists of a detailed synthesis of animal weaponry. Dr. Emlen outlines how competition for resources and mates can result in an "arms race" in which the individuals more likely to pass on their genes are the ones with the most impressive and oversized weapons. Illuminating modern examples of this can be seen in many insect and mammal species ranging from the harlequin beetle to the Irish elk. In what can be considered a runaway effect, certain species develop weapons that reach ridiculous sizes. At a certain point, Emlen points out that the arms race backfires on the individuals with the largest weapons, as less impressive individuals of the same species learn how to circumnavigate and ultimately out compete the previously dominant males. This, he argues, can explain the evolutionary changes and in some cases the extinction of many weaponized creatures. After setting the stage from a Darwinian perspective, Emlen expands his hypothesis to the realm of human weaponry and warfare, drawing parallels between the evolution of animal weapons and our own. He describes cases throughout human history where arms races have accelerated drastically and then become obsolete after a circumnavigation becomes possible by an opponent. His examples provide a convincing argument that maybe even our own technologies evolve under a mechanism similar to (or the same) as evolution by natural selection.

However, he recognizes that we have reached a point in human history where the historical trend may not hold. We now live in a world with weapons of mass destruction—nuclear, biological, and chemical weapons that hold the potential to wipe out our entire species many times over.

Waterways

The waterways of planet Earth create beautiful art. Over tens of thousands of years, rivers meander back and forth, cover their tracks, and are continuously creating new patterns. One thing especially fascinating about these these patterns is that it's not the goal of a waterway to create something beautiful. Rivers, streams, and lakes simply exist according to the laws of nature and the product is incredible. It serves as a good reminder to stop living as if we have a destination. To stop ruminating on the past, resisting the occurrences of the present, and worrying about what will unravel in the future. The past happened and cannot be altered, the present is here and unfolding, and the future is not a real moment until it becomes the present. Live according to nature, allow yourself to be mindfully carried downstream, and perhaps you will create something amazing.

A thin chance

If it were possible to get in your car and drive straight up into the atmosphere (I sometimes wish I could do this), you would be in outer space in less than an hour. If you could drill down in the opposite direction, you would be in the Earth's HOT upper mantle in less than 20 minutes.

Space is defined as the "void that exists between celestial bodies", and although it has no absolute starting point, the commonly cited value for the atmosphere-space boundary is approximately 100 kilometers (62 mi) above the Earth's surface. This is one of those facts that I relearn, becoming momentarily terrified, and then go back to eating my oatmeal. You can say it about many natural phenomena, but the atmosphere is one of those things that our existence undeniably depends upon. This miniscule film of gas provides us with water, weather, warmth, and air, all while protecting us from the relentless dangers of space. With equivalent ease, we overlook the fact that the crust of Earth relative to its diameter is equivalent to the skin of an apple.

100 kilometers of an essential combination of gases separates us from the inhospitable void of interplanetary and intergalactic space. At the same time, 10-50 kilometers of rock separate us from the hellish inferno of Earth's interior. The atmosphere provides us with the familiar, almost deceivingly comforting blue skies during the daytime, when in reality our planet is one alone and enveloped in darkness. The lithosphere offers a relatively stable and hospitable surface to live upon when we are entitled to no such thing. Comprehending the size of our atmosphere and lithosphere should give us a new appreciation of them and underscore the importance of investigating and protecting them.

The galactic year

As planet Earth orbits the Sun, the Sun orbits the center of the galaxy. The latter less appreciated fact is significantly more impressive, both in terms of speed and distance. On its clockwise journey around the galactic center, the solar system travels at a staggering speed of 230 km/s (143 mi/s), or ~1/1300 of the speed of light. At this speed, it takes approximately 225 million terrestrial years for solar system to orbit once around the center of the Milky Way, a duration of time known as the *galactic year*.

Let's consider some fun thought experiments in the context of the galactic year. The first question that comes to my mind is, "What was happening on Earth one galactic year ago?" Another way to phrase this question is, "What did out planet look like the last time we occupied our current position within the Milky Way?" Operating on such time scales, we can be thankful for the geological record. At 225 Ma, we must nestle ourselves back into the Triassic Period, a time when the land masses of the planet were bound together in the vast continent known as Pangea. A unique climate and biota blanketed the Earth, and dinosaurs were just beginning to appear on the scene. ~76% of all marine and terrestrial species at the time were approaching their extinction at the end of the Triassic (~201 Ma), an event which is thought to have enabled the dinosaurs to become the dominant land animals on Earth.

In the past galactic year, mountains have been created and destroyed, oceans and seaways have flooded and dried, species have come and gone, climate has gone up and down, humans have appeared, civilizations have risen and fallen, and technology has exploded.

What will happen in the next galactic year? How about the next 15? Next 200?

Galactic oscillation

Could there be a link between our solar system's movement through the Milky Way Galaxy and mass extinctions on Earth? It is proposed by Dr. Michael Rampino that Earth's periodic passage through the galaxy's disk could initiate a series of events that ultimately lead to geological cataclysms and mass extinctions.

We live on a dynamic planet within a dynamic solar system within a dynamic galaxy. As our rocky planet orbits the Sun, the Sun and its planets orbit the center of the galaxy. Interestingly, the solar system orbits the galactic core in an oscillatory fashion, similar to how a horse on a carousel moves up and down as it orbits the center of the ride. To complete a wavelength of this oscillation takes about 60,000,000 years, meaning that our solar system crosses the galactic plane once approximately every 30 million years. Rampino & Stothers (1984) propose that increased meteorite bombardment, increased basaltic volcanic activity, and extinction events occur at approximately this same interval of every 30 million years, suggesting that there is a possible correlation between galactic astrodynamics and geological events on Earth.

Why would passing through the galactic plane result in increased meteorite impacts, volcanism, and subsequent extinction events? One possible explanation could be the distal Oort Cloud being gravitationally perturbed by the increased matter concentrated along the galactic plane. More mass along the plane could provide the gravity necessary to throw rocky Oort Cloud objects towards Earth, resulting in cataclysmic events on the surface. However, it was quickly calculated that there is not enough matter along the galactic plane to sufficiently jostle the Oort Cloud (Thaddeus and Chanan, 1985). What else could potentially explain this controversial correlation?

Rampino suggests dark matter. Enigmatic dark matter (discussed in a future entry) makes up 85 percent of the total mass of the Universe, yet we cannot observe it because it does not interact with electromagnetic radiation. However, dark matter has been confirmed by its gravitational effects, which act the same as ordinary matter. Coming back to our extinction theory, it turns out that dark matter (like normal matter) is concentrated along the galactic plane. It is argued that not only could dark matter provide the additional gravity necessary to disturb the Oort Cloud, but could also accumulate in Earth's core, providing adequate heat to cause the increases in global volcanism that some propose occurs every 30 million years.

"This new hypothesis links geologic events on Earth with the structure and dynamics of the Milky Way Galaxy" (Rampino, 2015). What a cool idea! I do not totally support this theory, as there is a huge amount of uncertainty in essentially all datasets involved. That said, regardless of whether this theory is entirely correct, it is a good reminder of the interconnectedness of the cosmos and represents a display of big thinking and creativity. The cosmos is saturated with mysteries that need to be unraveled, and I think we need to scale our imagination with the size of the problem.

Tektonikos

Relative to its radius, the Earth's rocky crust is proportional in thickness to that of the shell of an egg surrounding its interior (or, if you recall, the skin of an apple). This thin crust with its underlying solid upper mantle (together called the *lithosphere*) is in a state of constant movement. The plate tectonic theory states that the large, rocky slabs of Earth's lithosphere are mobile, and that their movement over millions (and billions) of years results in upper crustal deformation. The lithosphere is essentially floating atop an approximately 150 kilometer thick layer known as the asthenosphere (from Greek asthenés—"weak" + "sphere"). This layer is far less rigid than the overlying lithosphere due to its relatively high temperature and pressure and can be thought of as a lubricating oil that permits the movement of tectonic plates.

Although the asthenosphere facilitates lithospheric movement and subsequent deformation, there is still debate surrounding the mechanism that drives plate motion. Two major theories can be thought of broadly as *slab push* and *slab pull*. Slab push occurs due to convection within Earth's mantle pushing the plates in much the same way that hot air is deflected sideways when it reaches the ceiling. Slab pull occurs at subduction zones, where commonly cold, dense oceanic lithosphere is drawn downward into the mantle. Plate tectonics

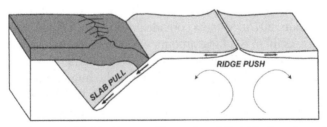

Schematic block diagram illustrating slab (ridge) push and slab pull, two possible driving mechanisms of plate tectonics on Earth.

is most likely driven by a combination of these two mechanisms, coupled with the viscous characteristics of the asthenosphere.

The convergence and extension of, as well as the lateral sliding of plates past one another, is largely responsible for the features we see on Earth's surface today. The lithosphere is brittlely and plastically deformed at the mercy of plate tectonics. Thankfully for geologists, plate tectonics shows the crust little mercy, and we get enormous mountains like the Himalaya and exploding volcanoes like Yellowstone and Mount Vesuvius. We get to investigate rocks that have been exhumed from the depths of the planet, carrying with them the secrets of these very mechanisms.

Flat slab subduction

When a relatively dense oceanic tectonic plate converges with a continental one, the former usually sinks (subducts) into the mantle. This subduction process commonly sends the oceanic plate down at an angle (~30°) and creates a predictable arrangement of geologic features at the surface (sedimentary basins, volcanoes, mountain ranges).

In some subduction zones, however, the subducting oceanic plate remains nearly horizonal and can migrate at a relatively shallow depth for hundreds or thousands of kilometers. It is well-established that this phenomenon, known broadly as *flat slab subduction*, exerts a first-order control on the tectonic and magmatic evolution of the overriding continent.

The western USA, with its impressive mountain ranges and magmatic rocks 1,000 kilometers from the modern-day plate boundary, offers a prime example of what a flat slab is capable of. Large mountain ranges exhumed along high-angle reverse faults (known as Laramide-style uplifts) and their

---

* Or for my more geologically savvy readers: a forearc, magmatic arc, a high-elevation hinterland orogenic plateau, retroarc fold-thrust belt, and a foreland basin system (e.g., DeCelles, 2004).

adjacent basins are partially a result of stresses being transferred from the Farallon oceanic plate during eastward-directed low-angle subduction. The timing of this subduction has been largely inferred from the timing and spatial distribution of inboard migrating magmatism (see Coney and Reynolds, 1977; my all-time favorite piece of scientific literature), but uncertainty remains surrounding the precise timing of when Laramide-style uplifts began. Modern analytical techniques (such as low-temperature thermochronology) are giving geologists insight into the cooling histories of the basement rocks that core these uplifts, in turn providing information on how flat slab subduction effects the upper plate (e.g., Carrapa et al., 2019; Ronemus et al., 2023; Caylor et al., 2023).

Examples of flat slab subduction occurring today exist in Chile-Argentina, Peru, Alaska, and Mexico. These regions offer geologists natural laboratories to understand how deep Earth geodynamic processes impact the geological architecture present at the surface.

Part IV: Cosmos

Dark matter

In the late 19th century, Lord Kelvin estimated the mass of the Milky Way using the observed velocities of stars orbiting its center. He calculated a mass far greater than the mass of observable stars, and stated that, "...many of our stars, perhaps a great majority of them, may be dark bodies." The first suggestion of what we now call *dark matter*. Other work in the early 1900s by Poincare, Kapteyn, Oort, Zwicky, etc. exposed discrepancies between the grouped mass of observed objects and their gravitational behavior. It was not until the 1970s that dark matter became recognized as a major unsolved problem in the astronomy community. Higher resolution work by Ken Freeman, Vera Rubin, and Kent Ford that utilized galaxy rotation curves showed that most galaxies analyzed require approximately six times more mass than is observed. Still, the problem was only mysteriously recognized in the appendices of scientific papers. "Additional matter which is undetected..." must be present (Freeman, 1970).

In the decades since, thousands more papers have been published that expose the necessity of dark matter to explain the evolution of galaxies, gravitational lensing, features in the cosmic microwave background, and the large-scale structure of the universe. Estimates suggest that dark matter accounts for ~85% of the mass of the universe (e.g., Carroll, 2007), yet it has still not been observed directly. It seems to be a form of matter that does not interact with the electromagnetic field; put simply, it is invisible. Most cosmologists think it is composed of an uncharacterized type of subatomic particle, non-baryonic but with the same gravitational effects that baryonic matter has.

Dark matter rests at the top of the list of unanswered questions about the cosmos; uncovering its secrets is an absolute necessity if our species wishes to understand our place in space.

Galaxy rotation curves

In general, the mass density of stars decreases as you move outward from the core of a galaxy. Kepler's Second Law—originally developed for planetary motion—uses this mass distribution to predict that the velocity of an object orbiting a concentrated mass will decrease systematically with distance from that mass. Rotation (or velocity) curves of disk galaxies plot the orbital speed of visible stars against their radial distance from the galactic center (see figure below). When the observed rotation curves from spiral galaxies are compared to curves derived from the distribution of luminous mass, a major discrepancy is revealed. Rather than a decrease away from the galactic core, we either observe a plateau or even an increase in the orbital velocity of stars. This discrepancy in rotation curves is one of the primary lines of evidence that point towards the existence of the non-baryonic mass known as *dark matter*.

Decades before the important work of Freeman, Rubin, and Ford (discussed above), Jan Hendrik Oort reported the first galaxy velocity measurements that were faster than the expected values based on visible matter. In the abstract of a 1940 paper, Oort stated that, "...the distribution of mass in the

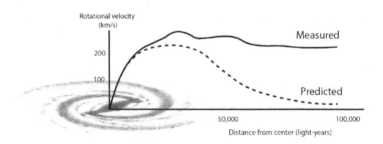

Measured rotation curve of a theoretical spiral galaxy (solid line) and a predicted one from distribution of the visible matter (dashed line). The discrepancy between the two curves can be accounted for by adding the effects of dark matter.

system appears to bear almost no relation to that of light." He didn't go so far as to suggest dark matter (because what a ridiculous idea haha!), but rather hypothesized that the discrepancy was due to extremely faint stars and/or interstellar gas and dust distributed throughout the outer galactic rim. It is now widely accepted that the galaxy rotation problem can be explained by the presence of non-baryonic cold dark matter, which despite being invisible, exerts the same gravitational effects as the visible matter that we are familiar with.

A mighty ring

"I have observed the most distant planet to have a triple form."
–Galileo Galilei, 1612.

When Galileo turned the first telescope to the night sky in 1610, "the most distant planet" known was the gas giant Saturn. Shortly before writing the sentence above, he recorded in his notes that Saturn appeared to have ears. He was confused when he observed the appearance and disappearance of these ears as the Earth made its orbit around the Sun. "I do not know what to say in a case so surprising...so novel," he remarked.

In the year 1656, the astronomer Christopher Huygens used a self-designed 50x refracting telescope (much more powerful than Galileo's) to observe the enigmatic shape of Saturn. In a publication three years later, he stated that, "[Saturn] is surrounded by a thin, flat, ring, nowhere touching, inclined to the ecliptic." An amazing discovery. Planetary rings? Such a thing was unheard of. Later, in 1675, Italian mathematician Giovanni Cassini first proposed that Saturn's ring was in reality composed of many smaller rings with gaps between them (a century later, work by Laplace proved that a solid ring was unstable and physically impossible). With the presence of rings around Saturn confirmed, astronomers realized that the disappearing ears that Galileo described was the result of the Earth passing in line with the plane of the rings.

Our understanding of Saturn's rings has come a long way. Composed primarily of small chunks of ice and rock, the rings are believed to consist of comets, asteroids, and potential moons that were torn apart by Saturn's strong gravity early in the solar system's formation (or more recently, according to new investigations; Iess et al., 2019, *Science*). High-resolution imagery shows us that rather than being a continuous disk, the rings contain many gaps, the largest of which have been opened by moons of Saturn that have cleared their orbital neighborhood. Many of the smaller gaps remain unexplained and are an active front of planetary science research.

A few of Galileo's first sketches of the gas giant Saturn, first in 1610 (left) and then with a slightly improved telescope in 1616 (right). Source: National Geographic Magazine.

The asteroid belt

"Between Mars and Jupiter, I place a planet." –Johannes Kepler, 1596.

The note above came from Kepler's analysis of Tycho Brahe's detailed observations of the architecture of the solar system. Kepler believed that the gap between the orbits of Mars and Jupiter was too large to be vacant. He was correct that the "vacant" space was occupied, but incorrect in his assumption that it was by an unrecognized planet. Instead, Kepler had made the first nod towards what we now refer to as the *asteroid belt*, a torus-shaped region consisting of hundreds of thousands of solid, irregularly shaped celestial bodies.

The asteroid belt formed out of the primordial solar nebula as gas and debris began to accrete into planetesimals (small precursors to protoplanets). These debris almost certainly would have continued to accrete to form a planet if it had not been for the great gravitational perturbation produced by the already formed gas-giant Jupiter. Jupiter's gravity resulted in the planetesimals having too much orbital energy for them to accrete into a planet; instead, the collisions resulted in a violent shattering and dissipation of material. Estimates predict that 99.9% of the asteroid belt's original mass was lost in the first 100 million years of the Solar System's history (Weidenschilling, 1977). Even with this much mass lost, the asteroid belt still contains a huge number of objects, with most recent estimates of the total asteroid count being ~1.1 million (>900,000 asteroids have been detected since the year 2000).

You might be wondering how we navigate our spacecraft (Voyager, Cassini, etc.) through such a huge amount of material without collision. The answer emphasizes a point that drives home the size of even the inner Solar System: the average distance between asteroids is >1 million kilometers (600,000 miles). This number is equal to about three times the distance between the Earth and the Moon.

Space is big.

Stellar evolution

The mass of a star determines the way it evolves over time. To gain a basic understanding of stellar evolution, it is helpful to separate stars into two separate categories: low mass and high mass.

Low mass stars (like our Sun) commonly become planetary nebulae at the ends of their lives. Once a star with less than 8 solar masses consumes all its hydrogen (the nuclear fuel that gives it energy) it expands to become a red giant. The bloated star then expels its outer layers, exposing a hot core. The hot core is called a white dwarf and ultraviolet radiation given off by it illuminates the discarded material. This illuminated material eventually entirely dissipates and leaves behind the white dwarf at its core.

Unlike our Sun, massive stars can fuse heavier elements than hydrogen and helium. A large star (>8 solar masses) will remain in stellar equilibrium—meaning that the energy generated by fusion reactions is sufficient to counter the force of gravity—until it reaches the fusion of iron. The fusion of iron produces no net energy output, so no further fusion can take place. The equilibrium of energy output and gravity is no more, and the star collapses in on itself at speeds up to 23% that of light. The temperature and pressure become so high that within seconds, a cataclysmic explosion expels the stellar material into space. This rapid collapse and violent explosion at the end of some high-mass stellar evolution is known as a Type II supernova. In the aftermath of this explosion lies either a rotating, extremely dense neutron star (pulsar) or the cosmic prison of light known as a black hole.

Nuclear fusion, exploding stars, and black holes! What a privilege it is to be alive today with an ability to learn about our place in space.

Exploding stars and tree rings

A study published in the International Journal of Astrobiology (Brakenridge, 2020) suggests that the timing and intensity of exploding high-mass stars may be preserved in tree rings on Earth.

Above we learned that when high-mass stars reach the fusion of iron, the equilibrium between energy output and gravity is lost. The star collapses in on itself and the temperatures and pressures become so great that the star explodes, violently ejecting its guts (plus lots of energy) into space. One of the primary types of energy ejected from these explosions takes the form of gamma radiation, which can trigger the formation of rare isotopes in the Earth system. Carbon-14 (radiocarbon) is an example of a relatively rare isotope in the Earth system that forms when the atmosphere is bombarded by cosmic radiation. Since the Earth is always being hit with cosmic radiation, Carbon-14 is always forming, but occasionally its concentration spikes dramatically. These massive spikes in the record that demand an explanation. Two main possibilities exist for such spikes: solar flares or distant supernova explosions.

Brackenridge (2020) investigates this question by comparing the timing of known supernova explosions to the timing of C-14 spikes recorded in Earth's tree rings. Eight of the closest supernovas in the study correlate to unexplained spikes in radiocarbon concentration, which the study concludes is strong evidence for a linkage between the two. One specific example is the explosion of a star in the Vela constellation approximately 13,000 years ago. Shortly after the star went supernova, radiocarbon levels went up by almost 3% on Earth.

There is still a lot of uncertainty in these findings, including error on the timing of stellar explosions and how to properly distinguish between Sun-related cosmic radiation and that from sources in deeper space. Regardless, it is probable that events in deep space play an important role in Earth's climate fluctuations and, by extension, the evolution of life.

Consider yourself a space child.

HR diagram

From their birth out of dense molecular clouds to their inevitable collapse and catastrophic explosion, the evolution of stars is awe-inspiring. Of course, the cosmos is filled with many different types of stars, each of which takes a unique path through stellar evolution. Stars come in many different sizes, undergo fusion at varying rates, reach our telescopes with diverse intensities of light, and burn across a vast range of temperatures. For these reasons, stellar classification represents a huge front of research in astronomy and astrophysics. It is almost impossible to discuss stellar evolution in these disciplines without an understanding of what is known as a Hertzsprung–Russell diagram (HRD).

I've looked at a lot of cool graphs in my days, but the HRD rests comfortably as my favorite of all time. Developed in 1910, the HRD is a plot showing the relationship between stellar classification and absolute magnitude. Put more simply, it displays the relationship between a star's temperature and its luminosity (brightness). On the x-axis is stellar classification, where stars are given a letter-classification based on their temperature (usually OBAFGKM*, with temperature decreasing to the right). On the y-axis is luminosity, which is a measure of

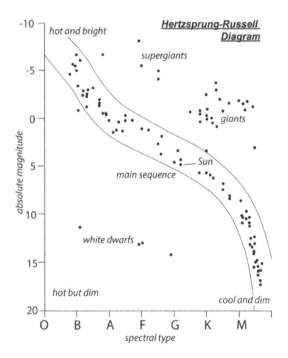

---

* My undergrad astrophysics professor suggested using the mnemonic "Oh Be A Fine Girl, Kiss Me" to remember the order of stellar types. Perhaps somewhat inappropriate in 2024 lol, but I've never once forgotten stellar classification.

the brightness (or energy output) of a star compared to that of our Sun. Confusingly, the lower the value for luminosity, the brighter the star. In general, larger stars like the red-supergiant Betelgeuse plot towards the top of the HRD.

There are three main regions on this diagram: (1) white dwarf stars plotted in lower left corner due to scorching temperature but small size, (2) red giant and supergiant stars in the upper third due to massive size but relatively low temperature, and (3) main sequence stars, which split these other two regions down the middle and represent stars that are hanging out during the hydrogen-phase of nuclear fusion.

The HRD plays a vital role in our understanding of stellar evolution and I think it represents one of the finest displays of astrophysical thought in the 20th century.

Aurora

During periods of solar unrest, the Sun violently ejects plasma-ionized particles towards its hosted planets. When these charged particles (mainly electrons and protons) reach planets that have magnetic fields, some of them become trapped along the planets magnetic field lines. Once the solar particles are trapped in a planets field lines, they travel to the poles and precipitate into the upper atmosphere, where they impart energy to gas molecules and excite them. The excited molecules of the upper atmosphere can essentially shake off the introduced energy and return to their normal state (i.e., drop from a higher to lower energy state). If you remember from your chemistry class, the "jumping down" of an electron from a higher to lower orbit is accompanied by the release of a photon (i.e., light). This process (more or less) is responsible for one of nature's most beautiful displays: the aurora.

As stated above, any planet harboring a magnetosphere can trap charged particles thrown to it by a coronal mass ejection. Consequently, we commonly see aurora on our celestial neighbors. The wavelength of the photons (color of light) released during the process outlined above is a function of how much energy is being exchanged and what type of gas molecule is being excited. A higher frequency photon (like blue) being released indicates that the difference between the normal and excited states of gas molecules in the upper atmosphere was relatively large. On Earth, the green glow of the aurora is due to the predominance of nitrogen and oxygen in our atmosphere. This contrasts to the blue-ultraviolet aurora on Jupiter, for example, whose upper atmosphere is dominated by hydrogen and helium.

There are many active research questions surrounding aurora, including why they take certain shapes and how they may affect the satellite communication and navigation systems of Earthlings.

Magnetars

Representing the collapsed core of a giant star, neutron stars commonly rotate about their axis around once every ten seconds and have exceptionally strong magnetic fields due to powerful electrical currents in their interior. They are also some of the densest objects in the known universe, as one tablespoon of substance from your average neutron star would have a mass of over 100 million tons. Good luck holding onto that spoon (also, imagine if you were able to get a bite into your mouth). As fascinating and ridiculous as these objects are, they have relatives that are even more impressive.

The magnetar is a type of neutron star that, as the name suggests, has an incredibly strong magnetic field. At $\sim 10^9$ to 10^{11} tesla, the magnetic field surrounding an average magnetar is several quadrillion times more powerful than the field surrounding planet Earth. Within their magnetic fields, photons are readily split in two or merge together, the vacuum of space itself becomes polarized, and your household hydrogen atoms are stretched into rods 200 times narrower than their normal diameter. All this is to say that the environment surrounding these star remnants is extreme. Scientists believe the incredibly strong magnetic fields result from a magnetohydrodynamic dynamo process (wow, fancy!) that is facilitated by their staggering density (Raynaud et al., 2020). To add to the chaos, magnetars rotate much faster than their neutron star cousins, turning over on their axis once every half-second. Stars on crack – they need to chill fr.

A personal thought—I find it mildly hilarious that I was able to read up on magnetars this morning over a double Americano. We humans casually read about the most ridiculous cosmic phenomena and subsequently move on with our lives.

Smooth pulsars

A study from the LIGO team (Abbott et al., 2020) has determined that pulsars (rapidly rotating and exceptionally dense neutron stars) are perfectly spherical and have completely smooth surfaces. This finding may not be surprising considering that neutron stars are held together by nuclear forces (rather than gravity), but the method the team used to determine the sphericity for various pulsars is fuc*ing radical.

LIGO (the Laser Interferometry Gravitational-Wave Observatory) can detect extremely minor distortions in the fabric spacetime. These distortions, known as gravitational waves, propagate through space from sources that perturb spacetime. The most well-known examples are merging supermassive black holes and colliding neutron stars (these events create relatively large gravity waves). However, smaller perturbations can now be observed using LIGO, such as in this study. The basic idea is that the amount of spacetime distortion (i.e., changes in the pulsar's local gravitational field) is related to how spherical it is. This makes sense intuitively—if you had a wacky, misshapen pulsar rotating 30 times per second, the gravity field would be getting thrown all over the place. This would result in the creation and propagation of gravity waves toward our telescopes. Conversely, imagine a perfectly spherical, perfectly smooth object rotating in space. Very little spacetime distortion, nice and neat.

This study measured the degree of gravitational wave propagation from five different pulsars. Results from all five showed no significant spacetime distortion. With the sensitivity of the instruments, this means that a pulsar 20 kilometers in width must be perfectly spherical to within ~50 microns. In other words, if a hair's breadth were added to one side of the neutron star, it would create detectable gravitational waves. Consequently, these objects must also be almost perfectly smooth. The ultimate cosmic billiard ball.

Dividing nature

"If our small minds, for some convenience, divide this universe into parts—physics, biology, geology, astronomy, psychology, and so on—remember that nature does not know it!"

This quote, modified from Richard Feynman's *Six Easy Pieces*, serves as a good reminder of the interconnectedness of the universe.

The universe is vast and complicated. It is necessary for researchers to focus most of their work within a given subdiscipline of an already narrowed-down field of inquiry. With millions of scientists worldwide that have varying interests and capabilities, our breaking down the universe into specified parts seems to work well for unraveling its secrets. Obtaining a deep enough understanding of Einstein's general relativity to theorize and detect gravitational waves cannot be done by geologists. Similarly, theoretical physicists do not have an adequate understanding of Earth processes to unravel the mechanisms of mountain building. As scientists and life-long learners, there is no other option than to direct our focus to a certain discipline of interest. That being said, we must always keep the interconnectedness of all events in the back of our scientific minds. As Dr. Feynman states so eloquently, our division of the universe into parts in unknown to the universe itself. The universe functions as one organism with all parts connected, and we must not forget that. The scientists who acknowledge that disciplines they don't fully understand influence their own field seem to be the ones who have the firmest grasp of the nature of reality. Today I'm going to take a moment and a mindful breath and recognize that my problems, although they seem separate from the universe, are part of it.

Nature is amoral

"Nature, who is as she is, and who existed in earthly form for 4.5 billion years before we arrived to impose our interpretations upon her, greets us with sublime indifference and no preference for accommodating our yearnings." –Stephen Jay Gould, *Rocks of Ages.*

This nice sentence raises several points that I have found helpful to consider in not only scientific work but also existential ponderings. First, it serves as a reminder of the great antiquity of our planet, a humbling reminder of the fact that Earth (and life for that matter) had existed for billions of years prior to our emergence. Secondly, "...to impose our interpretations upon her [the earth]..." braids nicely with the point Feynman emphasized above (our division of the universe into parts is unknown to the universe itself).

Arguably most important, Gould reminds us that nature is indifferent to our passions and desires and does not exist to accommodate our yearnings. This is not to say that nature is immoral, and although it can seem sometimes that nature goes out of its way to harm us, it is not the case. Nature is amoral, and simply exists according to the physical laws of the universe as it has for the past 13+ billion years. Nature is not on our side, but it is not against us either. It is up to us to work with it as well as we can to fix our mistakes and progress into the future.

Cosmic orientation

Are we trying too hard to orient ourselves in the cosmos? We are constantly simplifying, categorizing, and separating nature into parts that it does not recognize. Many lives are spent arranging long-term goals and trying to fulfill aspirations only in an attempt to feel ordered and satisfied. Could it be that these attempts at orientation are futile and taking away from our staggeringly unlikely conscious experience? I understand that the cosmos is way too complex to gain a whole knowledge of with our brief lives, and that our breaking down of the universe into specified parts seems to work well for unraveling some secrets. But I can also see too narrow of a focus stripping the cosmos of some of its true grandeur, leading potentially to a dull and incomplete perspective on reality. I guess in the end all we can do is maintain a perspective that keeps the interconnectedness of all events in the back of our scientific minds.

I have (perhaps backwardly) spent many of my days in awe of what we have learned about our place in space. As of late, however, I have been in almost constant awe of what we haven't learned—or of things that we might simply be overlooking or overcomplicating. Maybe we need more imagination. More creativity.

Does science converge on truth?

Seriously limited by our linguistic, intellectual, and imaginative capabilities, maybe we are operating under an illusion that our scientific efforts are getting us closer to what we seek.

I had a stimulating conversation recently on this topic, fueled by the acknowledgment that every interpretation we make about the cosmos is filtered through our limited human senses. We discussed the fact that we cannot escape our sensory limitations—even while using the scientific method. The systematic steps that we take in science are viewed by some as a way to get around our human limitations. In reality, everything in science—the hypotheses, the methods, the data, the interpretations, and the conclusions—are filtered through our biased and disillusioned mammalian brains (recall Don Hoffman's Interface Theory of Perception discussed in Pt. I).

116

As a possible analogy, we can consider the electromagnetic spectrum. As many of you probably know, the human eye is limited in its range of visibility, with most estimates pegging the visible light that we can see as 0.0035% of the entire electromagnetic spectrum. This means that our senses miss 99.997% of the universe's light. Why should this lack of visibility apply to only light? Why not everything else? It is likely that we are missing something crucial. Of course, we will not give up, but it should be kept in the back of our mind that our current efforts may be just scratching the surface. It's nice and idealistic to think of science as a gun that we are aiming at "truth" as a target. I think it is more likely that the target is not in sight and the gun has not yet been constructed. Maybe there isn't even a target?

Large numbers

When you spend a lot of time working with large numbers, it is easy to become indifferent to how large they really are. A trivial example of this can be found in considering the number of stars in our Milky Way galaxy, which is estimated to be around 250 billion. Our brains aren't capable of comprehending numbers this large by reading them, so scaling them down into more familiar units is helpful. Let's suppose that we were to dedicate an indefinite amount of time to naming all the stars in the Milky Way and that we were able to name one star per second. At this rate, it would take slightly less than 8,000 years to name the stars in our galaxy alone. Let's go deeper.

In the known universe, there could be as many as 2 trillion galaxies. Each of these galaxies can contain anywhere from 200 billion to 1 trillion stars. If you multiply the number of stars in our Milky Way Galaxy by the estimated number of galaxies in the observable universe, you get around 1 septillion stars. That is 1 followed by twenty-four zeros. Most scientists would agree that this is a gross underestimate of the actual number. Expanding on our thought-experiment above, if we were to name all the stars in the known universe, at one star per second, it would take ~32 quadrillion years.

The Hubble Ultra-Deep Field shown below puts things in perspective. Taken in 2004, this image spans one thirteen-millionth of the total area of the sky and contains an estimated 10,000 galaxies. Every single point of light besides a few obvious stars in the foreground represents a galaxy. Some of these galaxies are as far away as 13 billion light-years from Earth, and the light from their stars has traveled across the entire known universe.

These huge numbers serve as humbling reminders of the tininess of us and the enormity of the universe.

The famous Hubble Ultra-Deep Field image released in 2006. I've inverted colors from the original black and white image for clarity. Almost every point represents a galaxy.

Black Holes—they're simple

Black holes form by a relatively simple process.

I'm gonna review a few characteristics of big stars. A large star (>8 solar masses) will remain in stellar equilibrium—meaning that the energy generated by fusion reactions is sufficient to counter the force of gravity—until it reaches the fusion of iron. The fusion of iron produces no net energy output, so no further fusion can take place. The equilibrium of energy output and gravity is no longer, and the star collapses in on itself at speeds up to 23% that of light. The crushing weight of infalling matter compresses the star into a point of zero volume and infinite density called the singularity. During this compression, the temperature and pressure become so high that a cataclysmic explosion blows most of the remaining stellar material into space (Type II supernova, remember?). Left in the nebulous aftermath of this titanic explosion lies the singularity. A black hole is born.

According to Einstein's mathematics of general relativity, the infinite gravity at the singularity results in the breaking down of space and time as we know them, and the current laws of physics can no longer be applied. The mass and energy that approach the singularity lose dimensionality entirely. Since we cannot obtain data from beyond the event horizon, the gravitational singularity is theoretical and arises purely out of the math of general relativity. Its mathematical existence may even prove a flaw in Einstein's most brilliant theory and call for a united theory of quantum gravity. Quantum effects (physics at the smallest scale) likely become very important at the center of a black hole and general relatively is not compatible with quantum mechanics. A unified theory of quantum gravity may give us different answers to what lies behind the curtain of the event horizon, and it could be far stranger than we currently know.

Quantum foam

When solved for a black hole, Einstein's equations of general relativity suggest the existence of a region where spacetime is infinitely curved. This region, described to have a huge mass contained within zero volume, is known as the singularity. According to the laws of quantum mechanics, the existence of such a point is impossible. Thus, our two most accurate descriptions of the natural world are incompatible and therefore incorrect. It is for this reason that scientists are exploring the realm of quantum gravity, which may reconcile the differences between quantum and general relativity.

In his book *Black Holes and Time Warps*, physicist Kip Thorne speculates on quantum gravity, stating that near the singularity, "...[quantum gravity] unglues space and time from each other, and then destroys time as a concept and destroys the definiteness of space. Time ceases to exist. Space...becomes a random, probabilistic froth, like soapsuds." The "froth" that Thorne speaks of is what comprises the singularity in quantum gravity. No longer made of matter as we know it, the froth describes space not as any shape or curvature, but as a probabilistic space of curvature. Different curvatures have different probabilities of being present once quantum gravity takes over, and any curvature is permitted inside the singularity. Therefore, space and time no longer exist—they have been replaced by a probabilistic foam (what John Wheeler termed "quantum foam").

The illustrations below show three theoretical curvatures of quantum foam. The geometry of curvature within the singularity is probabilistic, meaning it might have a 0.3 percent probability for the form shown in (a), a 0.2 percent probability for (b), and a 0.001 probability for (c) and so on. Therefore, the only meaningful question that can be asked regarding the singularity is, "What is the probability it has the form (a), (b), (c), etc.?"

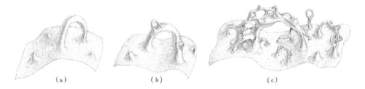

Embedding diagrams of quantum foam, illustrating that the geometry of spacetime on a Planck scale is probabilistic. Reproduced from Kip S. Thorne (W. W. Norton and Company, Inc., 1994).

L$_p$

The Planck length (L$_p$) is the smallest possible unit of measurement in the universe. At 1.6×10^{-35} meters, it is approximately 100 million trillion times smaller than the diameter of a proton. For perspective: a helium atom compared to the Planck length is comparable to the diameter of the Milky Way galaxy to the diameter of a bottlecap.

The Planck length is arrived at by combining three fundamental constants in a way that produces a length. These constants are Planck's constant (h; sometimes written as h/2pi), the gravitational constant (G), and the speed of light (c). In the order that they are written, these three constants are derived from quantum mechanics, gravity (Newtonian mechanics), and electromagnetism (the three major realms of physics). Put another way, the Planck length defines the meeting point of quantum mechanics, gravity, and time and space.

In our attempt to reconcile incompatible theories of the universe, we can compare these different realms of physics. We know that electromagnetism and quantum mechanics work well together (creating the field of relativistic quantum electrodynamics) and we know that there are serious problems when comparing general relativity (gravity) and quantum mechanics. The Planck scale provides us with the values that might represent the "breaking point" of all physics as we know it. The Planck length tells us that space itself is not properly defined on scales smaller than 1.6×10^{-35} m. At these scales, some physicists theorize that space and time become a random, probabilistic froth (quantum foam). It is possible that all four fundamental forces—gravity, electromagnetism, and the strong and weak forces—merge into one force. Mechanisms of quantum gravity and superstrings may also take over at the Planck scale.

$$L_p = \sqrt{\frac{hG}{c^3}}$$

The Schwarzschild radius

In late 1915, Albert Einstein published his general theory of relativity (GR), essentially a mathematical statement that space and time are two non-separate aspects of spacetime. It took only a few months for Karl Schwarzschild to find the first solutions to Einstein's equations (meaning that he found values for variables that satisfied the equations). Schwarzschild's solution described the gravitational field of a spherical body that eventually became known as a black hole. These initial solutions were questioned by the greatest minds of the 1930's, including Einstein, because they suggested the existence of a "singularity"—a place where GR (and all the other known laws of physics) break down completely. This controversy remained until the mid-1960's. Although still not compatible with quantum mechanics, GR and Schwarzschild's solutions are now widely accepted to explain the formation, geometry, and evolution of black holes.

One of Schwarzschild's most important solutions predicts that any object with mass has the potential to become a black hole. This idea is most effectively explained with a brief discussion of what is known as a *Schwarzschild radius.*

When a non-rotating and electrically neutral body (a star) suffers gravitational collapse, it will eventually collapse to a size in which its own gravitational pull is great enough to retain light itself. This critical "size" is what is known as the Schwarzschild radius. Theoretically, any mass can become a black hole if it collapses to the Schwarzschild radius other forces in nature would prevent it from happening). Using the simple equation on the next page, one can determine what this radius would be for any object of mass. A few theoretical examples may facilitate understanding. A mass equivalent to that of Mt. Everest would have a Schwarzschild radius of approximately 3.0×10^{13} meters, which is smaller than the size of a single atom. Planet Earth has a critical radius that is about the size of a peanut. If collapsed to these sizes, both objects would theoretically have an inevitable in-falling of matter and create an infinite curvature in space-time.

$$R_s = \frac{2GM}{c^2}$$

Hawking radiation

In 1974, Stephen Hawking published a one-page paper in the journal *Nature* that applied quantum field theory to black holes. The paper, titled *Black Hole Explosions?*, showed mathematically that black holes are not actually entirely black, but rather emit small amounts of thermal radiation. These emissions are known as Hawking radiation and they play an important role in the demise or "evaporation" of black holes.

For any black hole to dissipate energy in the form of Hawking radiation, its temperature must be greater than that of the cosmic microwave background (CMB), which has a T of approximately 2.7 Kelvin. If this is not the case, a black hole will gain mass as it absorbs heat energy from the surrounding space. Black holes with a mass equivalent to one sun have a Hawking temperature of 62 nanokelvins, which is significantly less than what is required to evaporate. For this reason, any black hole larger than one solar mass (given to current temperature of the CMB) will continue to grow. To have a temperature >2.7 K, a black hole needs to have a mass less than the Moon—it is easy to apply the Schwarzschild radius equation to determine that a black hole of this mass would have a diameter less than a tenth of a millimeter (i.e., tiny black hole!).

All this is to say that the amount of Hawking radiation emitted from a black hole is inversely proportional to its mass. This means that bigger black holes emit less radiation than small ones. Thus, black holes of very small mass experience extreme dissipation as a result Hawking radiation. For example, a black hole with a mass of one metric ton would take one nanosecond to evaporate (important to note that this mass corresponds to a diameter of 10×10^{24} meters). Hawking's equations can be used to determine the evaporation time for supermassive black holes but can only apply to the distant cosmic future when the temperature of CMB drops below that of the black hole. These results provide necessary timescales that are enormous—it would take a 100 billion solar mass black hole 2×10^{100} years to evaporate (twenty duotrigintillion years).

Black hole information paradox

"[Does the universe evolve in an arbitrary way or is it deterministic?] The classical view, put forward by Laplace, was that the future motion of particles was completely determined, if one knew their positions and speeds at one time. This view had to be modified, when Heisenberg put forward his Uncertainty Principle, which said that one could not know both the position, and the speed, accurately. However, it was still possible to predict one combination of position and speed. But even this limited predictability disappeared, when the effects of black holes were taken into account. The loss of particles and information down black holes meant that the particles that came out were random. One could calculate probabilities, but one could not make any definite predictions. Thus, the future of the universe is not completely determined by the laws of science, and its present state, as Laplace thought. God still has a few tricks up his sleeve." –Stephen Hawking, 1999.

The loss of information that Hawking speaks of in the context black holes has become known as the *information loss paradox*. This paradox, resulting from a combination of general relativity and quantum mechanics, points out that all physical objects consumed by a black hole devolve into the same state. In other words, no matter what information you send beyond the event horizon, the only thing you ever get out (in the form of Hawking Radiation) is mass, charge, and angular momentum. Any other information is lost forever. A true loss of information beyond the event horizon violates fundamental components of quantum mechanics, specifically the property of unitarity, which in essence implies that information (in the quantum sense) is always conserved. As is clear in the passage above, Hawking believed that the forever disappearance of information beyond the event horizon discounts the idea that we live in a deterministic universe—a statement with implications for human free will and the broader evolution of the cosmos.

Quantum vs. classical mechanics

Richard Feynman said, "If you think you understand quantum mechanics, you don't understand quantum mechanics." This quote makes me feel good about myself, because I don't understand quantum mechanics. Quantum mechanics (QM) is a difficult and counter-intuitive beast of a theory, which is inconvenient because it provides some of the most profound insights into the nature of the universe. In trying to understand it, I have found a comparison to the preceding paradigm in physics (classical mechanics) to be helpful.

In classical mechanics (CM), objects are characterized as having a location and a velocity. It states that if you theoretically know the location and velocity of everything in the universe, you therefore know what everything is going to do. Simply put, QM takes classical mechanics and adds a set of additional rules. Without getting into the details of quantum state (or wave function) and the equation that governs what the quantum state does (Schrödinger equation), we can point to this additional set of rules as a set that arises when we observe a system. From a CM perspective, nothing changes when you observe or measure what a system is doing. In contrast, QM states that there is something fundamental about how we observe or measure a system (i.e., the state of a system drastically changes when we look at it). One example of this is how an electron orbits the nucleus of an atom unobserved (in a cloud, or wave function) and what we see when we observe it (as a particle, with a position in space). This consideration of observation is one of the primary differences between the two theories.

I have seen the difference between QM and CM explained as the difference between a ramp and a staircase. In CM, events are continuous and move in a smooth and predictable pattern (i.e., ramp). In QM, terms like "quantum jumps" emphasize the unpredictability of the subatomic world, where electrons can make random transitions between two states and/or exist in two places at the same time (hence a staircase analogy, where particles discontinuously jump from one place to another).

I've also been intrigued by the von Neumann-Wigner interpretation of QM, which claims that consciousness itself plays a fundamental physical role in the collapse of the wave-function (broadly speaking, what happens when we observe a system). This interpretation essentially claims that consciousness is a necessary precondition that allows for the wave function to collapse and, therefore, for us to make quantum measurements. It is argued that our knowledge of any quantum mechanical system is based on its entrance into our consciousness. This interpretation requires consciousness to be separate from other physical processes and is criticized by most physicists as being dualistic.

Quantum mechanics

The quantum superposition principle states that a system is in all possible states at the same time—until it is observed. In other words, until it is measured, an atom can exist in multiple energy states and in multiple positions at the same time. Reread that sentence. Damn! There are many experiments and metaphors that can be used to help better understand the intimidating and enigmatic world of quantum mechanics. For superposition, I recently came across one that helped.

Imagine flipping a coin and letting it fall to the ground. In common experience you will look down once the coin has come to rest and see a definite result: heads or tails. Even if you don't look down, you can assume with 99.999% confidence it must be a heads or tail (I guess coins can land on their edge?). In the quantum world, things are different and mildly unsettling—the coin exists as both heads and tails at the same time. Quantum material properties do not exist until they are measured, so once we look, the material will take a form (e.g., heads). Until then, the material is literally in two states at once.

• • •

The universe is strange. Stranger than we can conceive, I'm convinced. Quantum mechanics is one of those examples of science, mathematics, and human imagination taken to its extreme. It doesn't make sense to almost all of us (myself included) because it flies in the face of our intuition and everyday observation. Whenever I attempt to revisit the quantum world, I always reconsider the linguistic and imaginative limitations that are probably keeping us from understanding true reality. There is no reason to think that our fragile and limited brains can comprehend what is happening in this strange cosmos. Maybe we will never know, maybe we will. We might as well try to understand.

Special relativity

Newton's famous Second Law, stating that the acceleration of an object is dependent on the net force applied to it and the objects mass (a=F/m), was believed to be true for the 200 years after its formulation. Albert Einstein discovered both the mathematical error in the Newtonian law of motion and its correction in the year 1905.

Newton's Second Law held the assumption that the mass of an object remains constant during acceleration—an assumption that is now known to be incorrect. The math of Einstein's Principle of Special Relativity reveals that the mass of an object actually increases with velocity, which is displayed in his eloquent equation shown on the following page. The correction to the mass of a moving object from Newton's static-mass is all you need to understand to have a fundamental grasp on special relativity.

A few things about this fascinating equation! First off, you may have noticed that the mass change must be small in ordinary circumstances, because it relies entirely on the speed of the object *relative (key word hehe) to the speed of light. At 300,000 km/s (186,000 mi/s), the speed of light serves as a worthy denominator. For example, at Interstate speeds of 145 km/hr (80 mph), the correction to mass is only one part in one hundred billion (impossible to observe). However, as the speed of objects with finite mass obtain higher speeds, the correction becomes more serious and more important. If an object with a resting finite mass of 1kg were accelerated to 90% the speed of light, its mass increases to ~2.3kg. At 99% of c, 7.1kg. What happens next is what I find most fun to think about!

Velocity (v) hits the speed of light and the denominator for the mass correction equation becomes the square root of zero, or zero, right? WRONG. Haha, you can't divide by zero homie! This is one of numerous ways it is possible for physicists to prove that nothing can reach the speed of light. It is truly a cosmic speed limit—one that we can get as close to as we want but will never reach.

$$m = \frac{m_o}{\sqrt{v^2 - c^2}}$$

General relativity

The relationship between gravity and time is a difficult one to grasp. However, once you get the basic idea it is one of the most amazing and thought-provoking relationships in the universe.

Broadly speaking, Einstein's general and special theories of relativity state that an object of mass within the universe creates a warping in the fabric of spacetime and that the speed of light is constant throughout the universe, respectively. Both theories must be applied to gain a basic understanding of the relationship between gravity and time. Since spacetime around a black hole is warped so severely (lots of mass), the distance that a beam of light needs to travel is greater near a black hole than it is around objects with lesser mass that do not distort spacetime as much. Here, special relatively becomes important. Since the speed of light is a universal constant, an observer in any given gravitational field must see it moving at its normal speed of 300,000 km/sec. The time/distance relationship of speed tells us that time is equal to the distance divided by the speed. Since here the speed remains constant, the time must increase proportional to the increase in distance (warpage of spacetime). In essence, general relativity (GR) tells us that if we have two clocks, each located in a different strength gravitational field, the clock that is within the stronger gravitational field will move more slowly relative to the other.

This is not mere theory or speculation. Einstein's mathematics of GR have been experimentally confirmed many times. One can even apply the elegant equations of GR to everyday life on Earth. A study by Uggerhøj et al. (2016) uses Einstein's mathematics to illustrate the effect of gravitational time dilation on the core and surface of the Earth. The authors provide calculations showing that over the course of our planet's 4.5-billion-year history, the pull of gravity has caused the core to be approximately 2.5 years younger than the crust. The authors initially calculate the value assuming a homogenous Earth, which of course is not realistic, but go on to employ density models that account for the differentiated structure of the planet, ultimately coming to their final age difference of 2.49 years. They also apply the same calculations for the Sun, showing that the age difference between its core and surface is upwards of 39,000 years.

Much to my delight, the authors conclude the paper by emphasizing the importance of fact-checking the work of prominent scientists. This idea of time dilation applied to the Earth system was originally proposed by the legendary physicist

Richard Feynman, who had reported wildly inaccurate estimates (which were subsequently cited by other prominent researchers). The authors state that, "In science, one route to becoming famous is being right on some important topics. However, just because someone has become famous, this person is evidently not necessarily right on all matters."

General relativity #2

If you were to approach a black hole, the flow of time would progressively slow down relative to where you started. This is commonly referred to as gravitational time dilation—greater gravity equals greater time elapsed. Since black holes have such incredibly strong gravitational fields, the effects of gravitational time dilation are most pronounced near their surface. One of the most confusing aspects of general relativity and time dilation is that the slowing of time only occurs relative to an outside observer. A helpful thought-experiment follows:

Imagine you are a future astronaut exploring the extreme gravity environment of a black hole. You design an experiment that consists of sending a clock from a set location all the way to the event horizon (gravitational point of no return). As the clock approaches the black hole, you observe that the clock appears to be moving slower and slower. When the clock reaches the event horizon, the stretching of spacetime and consequent time dilation becomes so severe that the clock stops entirely. This is a result of time getting close to infinitely slow at the event horizon. The confusing part is that all this time dilation only occurs from your observation point. If you were to approach the black hole with the clock, time would run smoothly, and you would not notice a slowing of time, movement, etc. Hence the relative in general relativity.

Black hole geometry

"The thing that has gripped me about black holes in the same way as black holes grip by their gravity anything that falls in their vicinity, is

the fact that a black hole is an object that is made not from matter, but from warped space and warped time." -Dr. Kip Thorne.

In the late 1600's, Sir Isaac Newton implicitly assumed that space and time were flat and independent of one another in his math of gravitation. Two centuries later, Einstein concluded not only that space and time are interrelated, but that they can be stretched, pulled, warped, pushed, and pulled by the very matter that they host. One of my favorite simple descriptions of general relativity is as follows: *matter tells spacetime how to curve, spacetime tells matter how to move.* In other words, the reason that the Earth orbits the Sun is not because of an invisible attractive force between the two, but because the Earth is travelling along a deformed surface created by the Sun. The deformed surface is spacetime. Extremely deformed spacetime is a black hole. Indeed, a common misconception surrounding black holes is that they are made from matter, when in reality they are made from warped space and warped time.

Let us consider black hole geometry. Suppose that you measured the circumference of a black hole to be approximately 60 kilometers. Using simple Euclidean geometry, we could easily calculate the expected diameter of approximately 20 kilometers ($d=60/\pi$). This is not what is found. With a hypothetical circumference of 60 kilometers, the math of general relativity shows that the corresponding diameter can be millions or even billions of kilometers(!). I read recently that the use of spatial distance can be misleading beyond the event horizon and that, depending on what the singularity actually is, the diameter of black holes may be infinite(!!). How can this be? We all learn at a young age the foolproof equations that describe the geometrical world around us—are they inapplicable here? As Lil Baby would say, YES INDEED! Space is significantly warped and consequently the laws of Euclidean geometry break down.

Coming back to our original question: With no matter, how does the warping of spacetime create the most massive objects in the known universe? It's here where the wonderous special theory of relativity comes back into sight. That is, mass is a highly compact form of energy, and the famous equation $E=mc^2$ tells us that they are interchangeable. The mass (energy) of a black hole is not concentrated in the singularity, it is concentrated in the warping of space and time. The initial energy (star collapse), plus the added mass of objects that fall into black holes, are stored in the fabric of distorted spacetime.

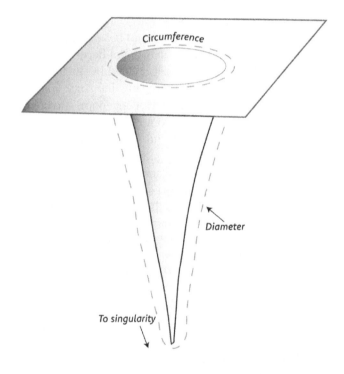

Illustrating the breakdown of Euclidean geometry in proximity to black holes. Modified from Thorne (1994).

Peculiar characteristics

I've read about some peculiar characteristics of the event horizon of a black hole. The event horizon, as many of you may know, represents the gravitational point of no return. Once past the event horizon, the gravitational pull of a black hole reaches a strength that is inescapable, and not even light can reemerge from the prison of the singularity. Another way of thinking about the event horizon is as the finite point in space where the escape velocity is exactly equal to the speed of light. Some interesting geometrical properties arise from this point in space. To an outside observer, the horizon is believed to appear as a static, unmoving spherical surface (as portrayed in images). In reality, the event horizon is actually moving outward at the speed of light, and it is this outward movement that does not allow light to escape. Since the horizon is moving away from the singularity at the speed of light, it would be necessary to exceed light-speed to escape, which as far as we know is impossible. The motion of the event horizon may not make sense, and that is because it doesn't in our normal understanding of how things move through space. The movement of the horizon is like Alice having to run as fast as she can just to remain stationary in *Through the Looking-Glass*.

Once past the event horizon, all matter is doomed to fall towards the singularity. The extreme distortion of spacetime near the center of a black hole has interesting consequences regarding general relativistic time. As the radial distance (r) from the singularity decreases, Einstein's math tells us that the time coordinate (t) grows larger and larger (as $r\downarrow$, $t\uparrow$). Eventually, all matter inevitably reaches the singularity ($r=0$), theoretically sending time to infinity. Time stops.

Gravity waves

Gravitational waves are ripples in the fabric of space-time, created by the most energetic and violent processes in the Universe. In 1915, Einstein published general relativity and

133

demonstrated that massive accelerating objects like orbiting black holes would disrupt space-time in such a way that "waves" of distorted space would radiate from the source. These waves propagate through the Universe at the speed of light and carry with them information about their origin, as well as clues into the nature of gravity itself.

The events that generate gravity waves involve big mass and energy conversion, yet it is still very difficult to detect them. For example, when the Laser Interferometer Gravitational-Wave Observatory (LIGO) first detected waves given off by black hole merger, the ripple in spacetime changed the length of a 4-kilometer interferometer arm by 1/1000 the width of a proton. The reason this distortion is so minuscule is due to the astronomical distances that separate our detectors from these events. The equation below shows the characteristic amplitude (height) formula for gravitational waves. As you can see, amplitude is inversely proportional to distance, meaning most simply that as you increase distance, the amplitude becomes really small. Since black hole mergers usually occur greater than a billion light-years away, the amplitude is reduced to a value that is extremely difficult to detect.

Our observation of gravitational waves may not be restricted to interferometers on Earth. A paper published in Physical Review Letters (Graham et al., 2020) is the first to recognize what is hypothesized to be light emitted from a black hole merger. The paper describes a black hole merger that sent massive amounts of energy through material of the accretionary disk, resulting in the stimulation of gas and dust molecules and the emission of photons. The merger, occurring in quasar J1249-3449, was detected by LIGO and simultaneously observed optically at the Palomar Observatory. Other possibilities exist for the cause of the light flare, but it is exciting that we may soon be able to observe such events visually.

$$h_c = \frac{16\pi^2}{c^4}\frac{f^2}{D}Ie$$

The small and the large

Scientists commonly use an understanding of the very small to gain insight into the nature of the very large. A technique known as *quantum cooling* allows for experimental apparatuses to be cooled down to temperatures extremely close to 0-kelvin (absolute zero). One way that this cooling is achieved, which can bring temperatures as low as 2 mK (0.002 K), is through the mixing of Helium-3 and Helium-4 isotopes. Due to quantum mechanical effects (specifically the Heisenberg uncertainty principle), the mixing of these isotopes results in an absorption of heat—ultimately allowing for the cooling of an apparatus or detector of interest.

Quantum engineers at Leiden University in the Netherlands have started using the technique outlined above (known as dilution refrigeration) to cool down a new kind of gravitational wave detector. Gravitational waves, as described in the previous entry, are ripples in the fabric of spacetime that propagate outwards from massive objects undergoing acceleration—the most common measurable sources being rotating neutron stars and colliding black holes. The detector is known as MiniGRAIL and consists of a 1.5-ton copper-aluminum spherical antenna. When cooled close to absolute zero, the atoms comprising this sphere come close to standing still, allowing for scientists to measure miniscule displacements of its spherical space (i.e., the kinds of displacements produced by the passing of a gravitational wave). In other words, small changes in the shape of the sphere occur during the passing-by of a gravity wave. The shrinking and expanding that occur are on the order of 10^{-20} meters, which is less than billionth the size of an atom, smaller than the nucleus itself. If the sphere were not cooled down to almost 0-Kelvin, it would have thermal vibrations of its own that are much larger than the vibrations produced by a passing gravitational wave.

The MiniGRAIL instrument is one of the many ways that quantum cooling and the science of the small can be used to investigate the large.

General relativity and quantum mechanics

Our two most fundamental and accurate explanations of the universe (general relativity and quantum mechanics, GR & QM, respectively) are not in agreement and cannot be reconciled. This is a big problem.

There are several places where GR and QM produce conflicting results, one of which being in the extreme gravity environment of a black hole. I encountered an interesting thought experiment that emphasizes an area of disagreement between our theories. I have adapted it for brevity. Consider two astronauts (Jon & Ric) in orbit around a black hole. Jon decides that, for the sake of science, he will change his trajectory and plummet past the event horizon in hopes of exploring the singularity. Ric watches from a distance as Jon approaches the event horizon. In accordance with Einstein's GR, Ric sees Jon getting ever closer to the horizon, but never crossing due to the slowing of time. Meanwhile, Jon's reality is quite different—he has comfortably crossed the horizon without noticing and is feeling the early onset symptoms of spaghettification (more on this phenomenon later haha). This is where we run into quantum mechanical problems. A fundamental rule of QM is that quantum information cannot be copied (i.e., it cannot exist in two places at once). Einstein's math tells us that space and time are not separate entities and supports the phenomena of *time dilation* (slowing of time for an observer). In short, GR tells us Jon is in two places in once, and QM tells us that that is impossible.

In the first section of this book, we entertained the idea that perception may not even approximate objective reality. The potential discrepancy between human senses and physical reality has huge implications for our approach towards and understanding of the cosmos. The disagreement between our current dominant theories proves that there must be something more fundamental, something deeper. It's clear that space and time are not what they seem.

At one with the black hole

A black hole is maybe the most mysterious object in the cosmos. At the center of black holes lies the gravitational singularity—a point where matter is crushed into an infinitely small amount of space. At the singularity, the curvature of spacetime is theoretically infinite, and with a large amount of matter in an infinitely small volume, density also goes to infinity. According to the math of general relativity, the infinite gravity at the singularity results in the breaking down of space and time as we know them, and the current laws of physics can no longer be applied. The mass and energy that approach the singularity consequently lose dimensionality entirely. With governing physical laws that are entirely different, or maybe even no laws (?), I think what lies at the center of a black hole could be literally beyond our wildest imagination.

In the classic book *Black Holes*, John Taylor describes the black hole nicely: "Inside this object (formed by the collapse of a heavy star to such a condensed state that nothing, not even light, can escape from its surface) the fundamental laws governing our universe appear to be destroyed, along with our usual concepts of space and time." He goes on to speak about the philosophical implications of such an object with the following wondrous passage:

"The black hole not only puts the scientific world in turmoil but also challenges many of man's basic ideas about his surroundings and his place in them. The implications of the black hole for man are as important, if not more so, than for science. They are especially relevant to man's attempt to grapple with the unknown and push his powers of reasoning to the ultimate when he tries to answer the basic questions of life and death, of animate and inanimate matter. The black hole brings us face to face with the mysteries of the world."

I still think that answering (or attempting to answer difficult questions) could be the reason we are here.

Taylor finishes the preface of his book with the following:

"...the black hole must cause a radical change in our understanding of many concepts so long cherished by man—immortality, reincarnation, dialectic, space, time, mind, and the Universe itself and certainty in that Universe. In total, the black hole requires a complete rethinking of our attitude to life. It leads to the liberating conclusion: in the beginning there was no beginning; in the future there will be no future. Man is at one with the black-hole Universe."

Through the wormhole

Wormholes are theoretical structures that link two or more separate points in spacetime and have been popularized as potential conduits for time travel. There have been many theoretical propositions for how wormholes could facilitate time travel. One of my favorites involves traveling through a traversable wormhole that exits into a region of space with a higher gravitational field than was present at the wormholes entrance (preferably an extreme gravity environment, like near a black hole). Time dilation as a consequence of increased gravity would lead to slower aging at the exit of the wormhole. Put another way, the exit of the wormhole would be "younger" than the entrance when observed from the outside (remember that the time dilation would not be detected by the traveler, as time is only warped for an outside observer). Therefore, two synchronized clocks would remain synchronized for the traveler. However, for an observer, the traveler would exit the younger end of the wormhole at a time when it is the same age as the "older" entrance, effectively going back in time.

Complicated wormholes

In 1997, the physicist Richard Holman provided what I think is one of the best descriptions of the wormhole: "Wormholes are solutions to the Einstein field equations for gravity that act as 'tunnels,' connecting points in space-time in such a way that the trip between the points through the wormhole could take much less time than the trip through normal space."

I've always liked this description because of how Holman points out the potential implications that such structures would have for time travel. He doesn't explicitly state how the "...trip between points..." would take less time, and it allows for the imagination to take over (I personally envision spacetime folded over atop itself like a taco, with a warped bridge connecting the two halves). Of course, when most of us think of wormholes, we think of time travel. We think of science-fiction films in which wormholes are held open to allow for faster than light travel through the cosmos. Clearly, the popular conception of wormholes is that they serve as shortcuts; that is, the distance needed to travel through the bridge connecting the halves of my taco would be less than the distance around the entire shell to reach the same point. On a cosmic scale, we can imagine cutting many light-years off a journey by using such a shortcut.

My conception of wormholes was challenged when I learned that, in terms of the theoretical physics of wormholes, there is no particular reason why the distance through them should be shorter. Perhaps the bridge connecting my taco halves is much longer than the assumed "long way round." Montana State University professor William Hiscock points out that wormholes may be analogous to a long, gnarled, and complicated hole rather than a straight shot. Indeed, some worm holes in apples, for example, are not simple and direct. This concept of meandering and complex wormholes also made me consider that perhaps there are wormholes that travel through vast amounts of spacetime but eventually exit out of their entrance! Maybe this is what black holes do?

I hope a human goes into a black hole someday.

Spaghettification

The amusingly named process known as *spaghettification* refers to the stretching of objects into elongated shapes whilst in a non-homogenous gravitational field. The process can be thought of as a most extreme tidal force, where one side of an object experiences a more intense gravitational pull than the other (Jupiter's moon Io always comes to mind, as tidal forces are responsible for its internal heating and intense volcanism). The tidal forces during spaghettification, however, are so strong that any object to experience it is completely disassembled.

If you fell into a black hole, spaghettification would stretch your body past it's elastic limits and snap you apart right above the hips. The same mechanism would continue to break down each half of your body, and over the course of a few seconds you would be reduced to a string of disconnected atoms plummeting towards the inescapable singularity.

Stephen Hawking famously described a fictional situation in which an astronaut approaches the event horizon of a black hole feet-first. Newtonian mathematics tell us that the force of gravity is inversely proportional to the square of the distance between two objects. Therefore, as a human body approaches a black hole, the difference in distance from the toes to singularity and the head to singularity would have disastrous consequences. The extreme, non-homogeneous gravitational field would provide a drastically different gravitational pull on the feet and head, resulting in spaghettification. When the body would reach its elastic limit and snap would depend on the size of the black hole consuming it (probably beyond the event horizon for a supermassive black hole). Regardless of when it occurred, the stretching effect would not stop after breaking you in two. It would continue to work on the resulting parts and ultimately break your body down into its smallest constituents. Therefore, our bodies and brains would arrive at the singularity represented by the atoms that composed them (or maybe even smaller components?). What happens to matter at this point is unknown and may be one of the unanswerable questions of the universe.

Temporal paradox

Is time travel possible? We have now discussed how it is theoretically possible to travel into the past by utilizing wormholes—bridges that connect disparate points in spacetime. Although general relativity can be applied to suggest that such time travel may be possible, a philosophical approach reveals paradoxes that may discredit the entire idea.

Temporal paradoxes are logical contradictions that arise when the past is altered in any way. The well-known "grandfather paradox" plays with the idea that if a time traveler were to go back into the past and kill their own grandfather, it would prevent their own conception and existence. Therefore, there would be no one in the future to have gone back in time! Also known as a "consistency paradox", this same idea can be applied to any action that alters the past. Since it is true that the past occurred in a certain way to lead to one's existence, it is a logically contradictory for the past to have occurred in any other way. Therefore, even if we were able to travel back in time using a wormhole, the traveler(s) would necessarily have to allow the past to play out as it had previously.

In *A Brief History of Time*, Stephen Hawking offers a possible resolution to this paradox. He states that everything that happens in spacetime must be a consistent solution to the laws of physics, meaning that you could not go back in time unless history showed that you had already arrived in the past and had not killed your grandfather (or committed any other act that alters the present). Essentially saying that if time travel is possible, it must be impossible to change recorded history. This conclusion has other interesting consequences for the concept of human free will.

Rephrasing the consistency paradox is helpful: *if* time travel actually is possible, it would be impossible for a traveler to change history since it has to repeat itself exactly. This means that if we are someday able to travel into the past, we will be restricted by a complete lack of free will. We would have to act in a way that led exactly to the future from which we left. If tomorrow a future human arrived on Earth, their future actions (along with everyone else's) are already recorded history to the civilization that they left. I may be beating a dead horse at this point, but mind-bending concepts like these require extensive examples to set in (I think).

We do not have to look so far as time travel to realize that free will is likely an illusion. The fact that you have arrived at this point in the essay at this exact time was not decided by you. I could have placed it in a different location, or you could have been distracted by a text message, either way your existence here reading is a direct result of external forces beyond your control.

Light year

The unit of distance known as the *light year* is used extensively in the fields of astronomy and astrophysics. As we discussed in Part I, it is easy to take things for granted when they are always there or consistently used. Sometimes when I can't sleep, I find myself thinking about how huge a light year is.

At 1.07 billion km/hr (671,000,000 mi/hr), light travels through our universe in all directions at constant speed. Since human minds do not prefer to deal with such large numbers, we created the unit of distance known as the light year—the distance that light travels in one calendar year. Some examples help to comprehend what this means. The closest star to Earth (other than the Sun) is Proxima Centauri, located approximately 4.22 light-years from Earth. This means that the moment a light-wave leaves Proxima Centauri, it will not arrive on Earth for 4.22 years. The light wave is traveling at the absurd speed of 1 billion km/hr and it still takes over four years to make the journey! Bruh.

Understanding the speed of light also has interesting implications with regards to the passage of time. In astrophysics, we most commonly discuss vast cosmic distances like millions or billions of light-years. Extending the explanation above, this means it takes light waves millions or billions of years to make the journey to our Earth-based telescopes. The sources of electromagnetic radiation (black holes, pulsars, quasars, globular clusters, etc.) that lie billions of light-years away no longer exist, as we are seeing them as they were billions of years ago. We are looking back in time—we are time traveling. Many of the objects we observe throughout the cosmos began their journey towards our telescopes millions to billions of years before our solar system had even formed!

Earth compared to the universe

From space, it becomes apparent that our world is but a lonely speck (a "mote of dust" as Carl Sagan famously said). A small blue spheroid spinning through the expanse of the cosmos. It can be difficult to recognize exactly how small our home planet is due to the awesome features it surrounds us with. Mountains and oceans are mighty and intimidating to our fragile human bodies, but on a cosmic scale become negligibly small. Let us consider the Earth relative to its surrounding space.

For this thought experiment, let's round the diameter of Earth up to 13,000 kilometers. Simple division tells us that Earth's diameter is equal to approximately 0.0085% of the distance to our Sun. Another simple calculation tells us that Earth's width makes up $6.7x10^5$ or 0.000067% of the solar system's diameter. A mote of dust indeed, but how about calculating the proportion of the universe that planet Earth occupies? Of course, all we can work with here is the observable universe, which has a diameter of approximately 93 billion light-years. Volume scales as radius cubed, so after converting to radius we can conclude the following: (radius of earth)^3/(radius of universe)^3 = $3.0*10^{-60}$. The Earth occupies 0.000000000000000000000000000000000000000000000000000000000 000000003% of the observable universe.

Bands, jets, and storms

A comprehensive theory of Jupiter's atmosphere has not been developed; perhaps not surprising given its complexity. The lack of a solid surface (the interior of Jupiter is fluid) makes it difficult to describe the atmosphere in terms of terrestrial meteorology, as convection may occur on enormous scales into the depths of the planet. Vasavada (2005) remarks that any comprehensive theory must explain (1) the existence of the stable bands that stripe across Jupiter's body, (2) the strong eastward jet (thin zones of high wind between bands) observed at the equator, and (3) the origin and persistence of large storms like the Great Red Spot.

Two broad classes of theories exist for Jovian dynamics: shallow and deep. The shallow models were first developed in the 1960s and suggest that the patterns we see at the surface are driven by small scale turbulence that is confined to the relatively thin upper layers of the atmosphere. These models are based largely on terrestrial dynamics and suggest that convection is related to the condensation and evaporation of water. The deep models suggest that the patterns are a result of deep-seated circulation that could possible reach down into the liquid core of Jupiter. Many of the deep models call upon the Taylor-Proudman theorem of fluid mechanics, which shows that fast-rotating liquids are commonly organized into a series of bands or cylinders parallel to the rotational axis (Busse, 1976).

Both models are good at explaining some things but have major pitfalls. The shallow models can explain smaller jets but do not explain the major equatorial jet. In addition, many features initially appear but do not persist. The deep models easily explain the equatorial jet and, in some cases, the bands, but we are still limited in our ability to produce the realistic 3D flow models required.

Jovian atmospheric dynamics are a major outstanding question in fluid mechanics and solar system astronomy. Most researchers think that an accurate model will incorporate aspects of both models.

Lightning on Jupiter

When a cloud acquires an excess of electrical charge (either positive or negative) and is able to break down the resistance of air, a visible discharge of electricity results. This discharge is what we know as *lightning*.

When positively charged particles at the bottom of a cloud create a channel of partially ionized air—in which neutral atoms have been converted to electrically charged ones—an initial lightning stroke can form and leave the cloud. When this initial stroke of electricity is about 30 meters (100 feet) from the ground, a connecting discharge of opposite polarity rises and meets it. When this is complete, a bright stroke of light propagates from the ground to the cloud at around one-third the speed of light. This flash of electricity can contain up to 100 million volts and peak temperatures of ~30,000°C—hotter than the surface of the Sun. These connecting discharges can occur within a cloud, between two or more clouds, or between the cloud and the ground.

It has long been known that lightning is not a phenomenon special to Earth. Other celestial bodies, most notably Jupiter, have shown intense electrical storms. Although Jupiter's lightning is like Earth's in many ways, there are several important differences. While most of the lightning on Earth occurs in the warm and moist mid-latitudes, on Jupiter it mostly occurs near the poles. Why is this the case? The answer to this long-lasting question was found by studying the distribution of heat in the Jovian atmosphere (e.g., Brown et al., 2018). Although Jupiter receives 25 times less sunlight than Earth, solar radiation is still able to warm its atmosphere. Brown et al. conclude that solar radiation creates stability in Jupiter's equatorial upper atmosphere, which inhibits the rise of warm air from within the planet. Since the poles do not receive as much warmth from solar radiation as the equator, warm air from the depths of the atmosphere is able rise in a convective fashion and facilitate the electrical storms observed at the surface.

Another study published in the journal Nature suggests that Jupiter also hosts a special type of electrical storm known as "shallow lightning". This research (Becker et al., 2020) is the first to propose that some of Jupiter's lightning originates from upper-atmospheric clouds containing a solution of ammonium and water (rather than just water, like on Earth). This study is intimately linked with other recent work from Guillot et al. (2020), which suggests some of Jupiter's violent storms form large "...slushy ammonia-rich hailstones...", which the authors refer to as *mushballs*. These mushballs are hypothesized to play an important role in the cycling of ammonia and distribution of electrical charge in the upper Jovian atmosphere.

Volcanoes on Io

With over 400 active volcanoes, Jupiter's moon Io is the most geologically active body in our solar system. Io's volcanism is a result of the moon's extreme internal heat, which is super high due to a phenomenon known as tidal heating. Tidal heating occurs when orbital energy is dissipated as heat in the crust and interior of a planet or moon. Differences in gravitational pull across a celestial object result in friction heating of its interior. Since Io is closely hosted by the most massive planet in our solar system, it experiences some of the strongest tidal forces of any moon. The side of Io that is furthest from Jupiter experiences a slightly smaller gravitational pull than the nearest side, resulting in a distortion of the moon's shape. With a relatively eccentric orbit, Io's distance from Jupiter changes significantly during one revolution. This change in distance is continually distorting Io's shape, leading to intense frictional heating of its interior and crust.

Many of the volcanoes on Io's surface are in a state of constant eruption, covering the surface with volcanic deposits and ejecting plumes of sulfur gas that rise hundreds of kilometers into space. It is a rather hellish moon, with most of its surface covered by vast plains of sulfur and sulfur-dioxide frost. As if Io weren't interesting enough geologically, it is also home to several mountains larger than Mt. Everest that rose because of compressive stresses in its crust.

The mountains of Io are incredibly impressive and the mechanism by which they form is poorly understood. The highest peak on Io is Boösaule Montes; at 17,500 meters (57,400 ft) it is twice the height of Mt. Everest (measured from sea level). The 115 named mountains on the moon have an average height of 6,300 m. With no evidence of global tectonic processes like those that create Earth's largest mountains, it has been proposed that these peaks form because of horizontal compression due to the mass contributed to the surface by continuous volcanism. Essentially a "squeezing out" of existing crust from beneath the massive weight of volcanics at the surface. There is a lot of research to be done on this problem—I am quite interested. And despite how shitty is sounds, I still want to visit.

Dunes on Titan

Electrified sand dunes, liquid methane seas, 600-meter deep canyons, and a significant atmosphere make Saturn's moon Titan one of our most interesting celestial neighbors.

Titan's dune fields are especially intriguing. Located along the equatorial region of the moon, the dunes reach heights of up to 100 meters (330 feet). And unlike dunes on Earth, which have a morphology predicted by the dominant wind direction, Titan's dunes do not. In fact, the dunes on Titan point in the opposite direction than expected. Climate simulations indicate surface winds that blow to the west—but the dunes point to the east! A conundrum.

Such strange dune behavior may be explained by the composition of sand particles on Titan and how they interact. Earth's rock (and consequently, its sand) is made of small silicate grains such as quartz. In contrast, Titan's "rock" is frozen water ice and frozen hydrocarbons (such as methane and ethane). The combination of this lower density detritus combined with cold temperatures (-179°C) and lower gravity turns out to be the perfect recipe for the building of immense electrostatic charge. This has led scientists to hypothesize that the backwards dune orientation on Titan is a result of the dunes being held together by electrical forces.

Mendez et al. (2017) conducted an experiment in which they tumbled complex hydrocarbon particles within a small tube containing pressurized nitrogen (the dominant gas of Titan's atmosphere). This experiment, designed to simulate the behavior of sand grains in Titan's atmosphere, showed that the electrostatic forces between hydrocarbons on Titan should be an order of magnitude greater than those between sand on Earth. Thus, it is plausible that the dune orientation (and perhaps their great height) can be explained by electrostatic attraction, like the force that holds a balloon to your skin after moving it back and forth.

Water on Mars

On Earth, liquid water laps against the shores of lakes and oceans, carves impressive canyons, transports mass amounts of sediment, and is responsible for the deposition of many of the rocks we see exposed at the surface. Earth is not the first celestial body to have been shaped by these forces. Approximately four billion years ago, long before they existed on Earth, these same processes sculpted the red planet Mars.

Evidence of previous liquid water on Mars is overwhelming. Ancient networks of river valleys, enormous flood-carved outburst channels, deltas, lakebed deposits, and ancient shorelines have all been extensively documented with rovers and satellite imagery. A particularly clear and interesting example of ancient water flow on Mars exists in the southern highlands, where a dendritic (tree-like) system of river valleys is carved into the surrounding topography. Less pronounced valleys branching off from a central channel represent ancient tributaries that ultimately contributed water and sediment to a massive depositional basin at a lower elevation. Similar dendritic river patterns exist all over planet Earth. The

Colorado River of the western USA is a textbook example that may be analogous to the channels that existed on the southern highlands of Mars. Another tantalizing question raised by the presence of fluvial processes on Mars, which would have required a much warmer and wetter climate, is whether the ancient environment could have supported life.

The prehistoric and present presence of water on Mars remains a frontier of planetary science and will continue to be studied by scientists and eventually astronauts. Let's go!

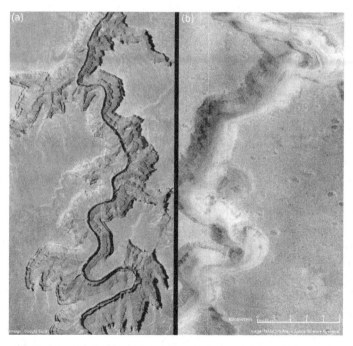

Similarities in channel morphologies on Earth and Mars (left and right, respectively). Imagery sourced from the National Aeronautics and Space Administration (NASA).

Valles Marineris

Located along the equator of Mars, the series of canyons known as Valles Marineris (VM) is more than 4,000 kilometers long, 200 kilometers wide, and up to 7 kilometers deep. Despite being the largest canyon system in the solar system, its origin remains uncertain. Geologists and planetary scientists have put forth many different hypotheses as to how it formed that can be broadly divided into (1) tectonic processes, (2) erosional processes, and (3) gravity-driven processes.

Tectonic processes that could be responsible for the formation of VM include rifting (the earliest explanation; e.g., Carr, 1974) and/or large-scale strike-slip faulting with resultant transtensional deformation of Martian crust (most recent hypothesis; Yin, 2012). The studies that attempt to link its formation to erosion are quite creative. One suggests that emplacement of massive dikes (molten rock) into preexisting crust caused melting of huge amounts of ground-ice and catastrophic flooding, which ultimately carved the canyons (McKenzie and Nimmo, 1999). Another proposes that subsurface removal of magma or dissolvable material such as carbonate led to surface collapse (e.g., Adams et al., 2009). A possible mechanism of deformation related to gravity-processes is spreading/collapse due to thickening creating a thermally and gravitationally unstable crust (e.g., Webb and Head, 2002). Of course, these causal mechanisms are not mutually exclusive—it is possible that all contributed to create VM.

I am especially intrigued by the gravitational and or thermal collapse model that involves an overthickened Martian crust. The image here shows the location of VM in relation to the Tharsis Rise, which is a vast volcanic plateau with an average elevation of 10 kilometers (33,000 feet) above the mean. VM clearly formed along the flanks of this region of crust that was significantly thickened by magmatic activity. On Earth, it is common for regions that experience extensive magmatism and crustal thickening to subsequently undergo large-magnitude extension (Coney & Harms, 1984; Howlett et al., 2021). I hypothesize that something similar is happening here.

There is so much unexplored territory in planetary science, and there is a need for more field-based geologists to apply some of their knowledge of Earth to other planets, especially Mars. If you are a young student, or exploring options for a university degree, I cannot recommend the geosciences enough. More generally, the people who are solving big, important questions in science are not much different from you...do not be deterred thinking that you must be a genius (I mean, look at me). Either way, STEM RULES!

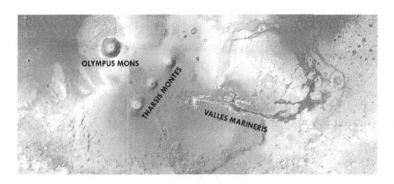

Slab-rollback on Mars?

Pole to pole, the crust of Mars is covered with interesting geologic features, some of which may be analogous to those on Earth. For me, the most interesting of these features are the Tharsis Montes volcanoes, located near the northwestern flank of the Tharsis rise (see figure above). The Tharsis rise is a vast volcanic plateau with an average elevation of >10 kilometers above the mean, traditionally thought to have grown as a result of a massive volcanic "hot spot" analogous to the one responsible for the formation of the Hawaiian Islands on Earth. However, this theory does not easily explain the geometrical configuration of the Tharsis Montes volcanoes, which are arranged in a linear fashion (trending northeast to southwest).

A more recent tectonic explanation was proposed by geologist An Yin that involves plate subduction and "roll-back" of a subducted slab beneath the Tharsis region (see Yin, 2012). In this model, Yin suggests that a massive meteorite impact during the period of Late Heavy Bombardment (ca. 4 billion years ago) initiated southeast directed plate subduction. Following subduction initiation, devolitization of the subducting tectonic plate is hypothesized to have caused the linear chain of Tharsis Montes volcanoes. This explanation draws heavily from processes on Earth, where the release of volatiles (things like water and carbon dioxide) during subduction causes partial melting of the overlying plate and can result in the creation of buoyant magma that rises to the surface. One of the ultimate surficial expressions of this subduction process is a volcanic arc—a chain of volcanoes that forms at the top of the overlying plate. Modern day examples of these linear volcanic belts exist in the Cascades of the northwestern USA and the Andes of South America (which look pretty similar to the belt of volcanoes on Mars).

I love this model but am unsatisfied by its treatment of Olympus Mons (OM), the gargantuan volcano existing adjacent to Tharsis Montes. With no evidence of a subduction zone interface (something resembling a trench) to the northwest of OM and the transition from an arc-like volcanic chain to a single volcano, it is more difficult to explain with a slab-rollback model. Additionally, OM is located predominantly aside the Tharsis Rise, isolated at a lower elevation. I am entertaining the hypothesis that arc-like volcanism caused crustal thickening, volcanic loading, and eventual downward bulging of the Tharsis Rise. This is turn could cause deflection of the magma flow from beneath without changing its source location. Just an idea! So fun to think about!

The largest mountain in the solar system

Olympus Mons, located in the western hemisphere of Mars, is an enormous shield volcano that is approximately 25 kilometers (16 miles) high, making it nearly three times as tall as Mount Everest (the tallest mountain on planet Earth). Also, the diameter of 624 kilometers (374 miles) makes the volcano about the same size as the state of Arizona. The extraordinary size of Olympus Mons is likely because Mars lacks mobile tectonic plates. Earlier we discussed Earth's tectonic plates that slide around on a mechanically weak subsurface layer known as the asthenosphere. A combination of seafloor spreading ridges and plate subduction keeps the plates of Earth consistently moving. As a result, when mantle-sourced plumes of molten rock ("hotspots") rise into the overlying mobile tectonic plate, the associated volcanism at the surface migrates over time (e.g., Hawaii). During the eruption of Olympus Mons, the crust of Mars remained fixed over a stationary hotspot, and the volcano continuously discharged lava until it reached an incredible height.

During difficult times on the planet Earth, I think a lot about other worlds. Along with reading science fiction, which takes me wayyy far from this planet, I daydream about our eventual landing, exploration, and colonization of the red planet. I love to think about the new frontiers of geologic research that will open once we can get teams on the ground. An entire planet unexplored by humans!! Are you kidding me?! There is nothing as exciting for scientists as investigating unexplored terrain. This is what attracts so many physical scientists to difficult areas like Tibet and the Himalaya, the high Andes, Antarctica, etc. As well as the elusive theoretical landscapes in quantum mechanics, mathematics, etc.

Mars is the next frontier, the most extreme adventure, and a most exciting prospect for scientific research. I am more optimistic than not that humanity will thrash its way through difficult times, prevail, and our science will propagate outwards from the Earth like ripples on a pond into which a stone is cast.

An unexplained perturbation

In 1821, French astronomer Alexis Bouvard used Newton's laws of motion and gravitation to predict the future positions of the planet Uranus in its orbit around the Sun. Observation of Uranus in the following decade revealed an orbit that deviated substantially from the predictions. Several early hypotheses were proposed to explain this wacky orbit, including a diminished effect of the Sun's gravity, observational error, or perhaps perturbation by a distant and undiscovered planet...

The final hypothesis turned out to be the correct one. A large, gaseous, royal blue marble of a planet lurked along the lonely outer edge of our solar system. This planet we now call Neptune. There is controversy as to who should get credit for Neptune's discovery, as multiple people were working on the Uranus problem simultaneously. Regardless, the approach was generally similar, consisting of solving what we would now call an inverse problem. These are problems in which the parameters of a model are deduced from observed data (in this case it was solely Uranus' orbital data). John Couch Adams began working on the problem when he was an undergraduate, hand calculating how the path of Uranus would be perturbed by assuming a mass and position for a more distant object. The calculated path can then be compared to observations and the characteristics of the unknown object can be adjusted in a way that reduced the error (a regression analysis; many iterations). Frenchman Urbain Le Verrier did the same calculations and was ultimately the person who convinced a few astronomers to turn their telescopes to the region of the sky where he predicted Neptune to exist. Both Adams and Le Verrier correctly predicted the location of Neptune to within one degree, and the gas giant was first observed on Sep. 24, 1846.

The winds of Neptune

The icy giant Neptune perpetually swims through the vacant space at the edge of our solar system. Located nearly three billion miles (4.5 billion kilometers) from the sun, it is easy to take this mysterious, cold, and dark world for granted.

My mind was thoroughly blown by an article that discussed the fantastic winds that are present on planet Neptune (Suomi et al., 1991). By measuring the motion of the large methane and hydrogen sulfide clouds across the upper atmosphere, researchers have recorded wind speeds on Neptune reaching greater than 1,300 miles per hour (2,200 km/h). At close to two times the speed of sound, these winds would be capable of stripping Earth's surface bare is a matter of seconds, with the mountains not lasting much longer. There is still some uncertainty regarding why these winds reach such amazing velocity, but it is probably a combination of low-internal friction due to the upper atmospheric temperature of negative 220°C and relatively high internal heating. Where this heat comes from is unknown. It's possible that it is left over from Neptune's formation, but the fact that the high heat flow is not seen in the similar planet of Uranus makes this conclusion problematic. Maybe it comes from somewhere else...maybe there is a little civilization inside of Neptune.

Isn't it ridiculous that this massive icy planet is always out there, sharing our star? I occasionally find myself thinking about how crazy it is that regardless of the complex events taking place on Earth, Neptune is still orbiting the Sun with its moons and supersonic winds. Simply existing in accordance with nature. Today, amid all your human problems, take a moment to recognize that, far in the outer reaches of our solar system, lurks the giant blue planet Neptune. A humbling thought.

Ice moons

Fractured and reflective icy shells separate our telescopes from the global subsurface oceans of Europa and Enceladus. These moons, hosted by Jupiter and Saturn, respectively, represent two of the most intriguing celestial bodies in the solar system. The presence of cryogeysers and ice tectonics suggests that both natural satellites have internal oceans separating their shells from solid internal cores. It is accepted that these oceans remain liquid because of tidal heating, which occurs as gravitational and orbital energy between a moon and its host is dissipated as heat (the same forces that cause Jupiter's hellish moon Io to maintain its volcanic activity). The combination of these geological processes and physical characteristics make these moons two of the most likely locations in the solar system for potential habitability. It is possible that life could lurk beneath the ice, either in the ocean itself or against the ocean floor in a hydrothermal vent environment. I am intrigued by the latter possibility.

Hydrothermal vents are seafloor fissures from which heated and mineral-rich water flows. Usually a result of seafloor volcanic activity, these vents can form "black smokers", creating diverse and productive biological environments fueled by the heat and complex chemicals in vent fluids. I have always considered a hydrothermal origin of life on Earth as being one of the most realistic ones, and there is no reason to think that either simple or complex life couldn't evolve in these environments elsewhere. One model that would allow for the existence of complex life on Europa addresses the problem that Europa's closed-ocean composition is not sufficiently oxidized. It involves irradiation and oxygenation of ice on the surface and transportation of oxygen-enriched crust into interior ocean via ice-tectonic processes (so cool). This oxygen transport mechanism would have allowed for European oceans to reach the same oxygen level as Earth's oceans in ~12 million years (Greenberg, 2010).

The habitability of these icy moons makes them a serious target for future space exploration. Do you think they could harbor life?

Part V: Beyond

Fermi paradox and the Great Filter

There is a contradiction between the high probability estimates for the existence of life outside Earth and the absence of evidence for it. This contradiction is known as the *Fermi paradox*, and follows the line of reasoning below:

1. There are ~100 billion stars in the Milky Way galaxy, many of which are similar in size and stage of evolution to the Sun.
2. Hundreds of millions of these stars likely host Earth-like planets, and if life evolves similar on these planets, some may have evolved intelligent life.
3. With civilizations that are perhaps millions of years more advanced than us, it is likely that they have developed some form of interstellar travel.

The famous Drake equation is a probability argument used to estimate the number of extraterrestrial civilizations in our galaxy and can pump out some huge numbers (due to the vast number of potentially habitable planets in the galaxy). When used by optimists, the equation produces numbers between 1,000 and 100,000,000 civilizations within our galaxy alone. So where are all those aliens? We have seen no sign of extraterrestrial life.

The economist Robin Hanson first proposed that our lack of evidence for ET intelligence supports the idea of a *Great Filter*, which essentially serves as a probability barrier for evolving species. He states that the absence of ET visitors suggests that the process of starting with a star and ending with galactic colonization is usually stopped somewhere along the path. Importantly, we don't know where the Great Filter exists, and for our own sake, we can only hope it exists early in the timeline (e.g., the production of reproductive molecules; RNA). Perhaps we are one of impossibly lucky organisms that snuck past. If the Great Filter lies ahead of us (perhaps in the form of nuclear destruction, pandemic, etc.), it is unlikely that we will succeed in colonization of the Milky Way. Following this line of reasoning has led philosophers such as Nick Bostrom to emphasize the importance that the search for ET intelligence, life on Mars, etc. finds nothing (if life exists within the solar system, the Great Filter lies ahead of us). In a 2008 essay, Bostrom states that, "...in the search for extraterrestrial life, no news is good news. It promises a potentially great future for humanity."

There are other possible explanations for the lack of alien visitation, ranging from ET life deliberately avoiding us (haha) to extraterrestrial intelligence being entirely different from ours (i.e., they are "too alien"; intraparticulate, perhaps?). Some theorists even entertain the idea that they are here already, unacknowledged, living amongst us...

Intra Particle Intelligence

The Search for Intra Particle Intelligence (SIPI) is a possible solution to the Fermi Paradox that searches for intelligent life at the subatomic scale. Proposed by Australian AI researcher Huge de Garis, the theory speculates that a hyperintelligent lifeform (one that long ago transcended biology) could potentially find ways to maximize their abilities at profoundly small scales ($<10 \times 10^{15}$m scale, utilizing the quantum mechanical properties of quarks and gluons for processing speeds a trillion trillion times faster than modern ones).

One of de Garis' conclusions encapsulates his idea nicely, stating the following: "The hyper intelligences that are billions of years older than we are in our universe (which is about 3 times older than our sun), have probably "downgraded" themselves to achieve hugely greater performance levels. Whole civilizations may be living inside volumes the size of nucleons or smaller."

What a cool idea. We spend so much time thinking about and searching for extraterrestrial intelligence elsewhere in the universe, but perhaps it is already here, a part of what we consider laws of physics. SIPI is simply a scaled version of the SETI project (Search for Extraterrestrial Intelligence)!

de Garis himself states that his idea is "...is a real paradigm shift away from looking for non-human intelligence in outer space, to looking for it in inner space." He asks the question of why elementary particles (such as quarks, gluons, etc.) are such "carbon copies" of each other and proposes that what we perceive as random fluctuations on the quantum/subquantum level may be the thought processes of ultra-miniaturized hyperintelligent beings.

If you are interested in these ideas, I recommend de Garis' essays titled *Femtotech: Computing at the Femtometer Scale Using Quarks and Gluons* and *Femtotech: Computing at the Femtometer Scale Using Quarks and Gluons.*

The simulation argument

Is it possible that your life is part of a computer simulation?

Our existence is awfully strange, and we are still trying ti determine the origins and future of our species. One of my academic heroes, Dr. Nick Bostrom, has popularized what has become known as the *simulation argument.* This concept argues that all of us are part of a single intricate computer simulation, designed and run by a species far more advanced than we currently are.

In a 2003 paper, Dr. Bostrom lays out the simulation argument as follows:

1. As a species progresses technologically, computing power grows enormously.
2. A species more advanced than our own may use their computing power to run simulations of their forebears.

3. The simulated people are conscious (us).
4. It follows that our minds may not belong to the original race but "...rather to people simulated by the advanced descendants of an original race."

Although this idea seems like one of science-fiction, I think it should be given a certain amount of credit. The unlikelihood of human existence, the difficulties we have explaining the emergence of life from the primordial soup, and the arising of consciousness—these are hard unknowns. As far as we can tell, the evolution of any lifeform is incredibly unlikely. We therefore need to statistically consider which is more likely: evolution by natural selection or the simulation argument. Another creation myth perhaps, but the conclusion that we are living in a simulation may not only be possible but plausible, and perhaps even probable.

A fine-tuned universe

The same physical laws operate across space and time and we have determined hundreds of physical constants that govern the behavior of matter. It seems that it's only possible for life to exist when these constants lie within an exceptionally narrow range. This argument forms the roots of the *fine-tuned Universe* proposition, which broadly states that if any of the dimensionless physical constants were even slightly different, it would not be possible for life (or possibly even matter) to evolve. Physical constants such as the size of the electric charge of the electron or the gravitational constant (G) seem to be finely adjusted to allow matter and ultimately life to evolve. The nuclear strong force is another well-known constant that, if only 2% stronger than it currently is, hydrogen would fuse into diprotons rather than helium (Davies, 2003). This minor change in a single constant would essentially make star formation impossible, which is a necessary precondition for human life. Examples like this one go on and on, and scientists and philosophers have spent time trying to explain how and why we live in a universe of such specific physical constants. Some theologians use the fine-tuning of the universe to argue for a divine creator, as they believe that the overall unlikelihood of life evolving is greater than that of an omnipotent god existing. Others consider what is known as the *anthropic principle*, which states that the observations of the universe must be compatible with the conscious life that observes it. The fundamental constants and age of the universe therefore become unremarkable, because human life and our consciousness are destined to arise in such a universe.

The fine-tuning argument gets at the core of human existence—why are things the way they are? I think we are beginning to scratch the surface of this question and should keep our minds open to all kinds of possible explanations. Simulation? Divine creation? Cosmic randomness? We have no shortage of things to contemplate.

The anthropic principle is worthy of consideration. Bostrom defines it as a consideration "...that any data we collect about the universe is filtered by the fact that, in order for it to be observable in the first place, it must be compatible with the conscious and sapient life that observes it." The anthropic principle (in many cases) is intimately linked to some form of the multiverse theory, as it would provide a possibly infinite number of other universes in which fine-tuned conscious life wouldn't exist. Our next discussion will be on the relationship between the multiverse and the anthropic principle.

The anthropic principle

The physical constants that control our universe appear to be fine-tuned for the conscious life within it. The anthropic principle explains the fine-tuning argument (FTA) by stating that the universe must be compatible with the conscious life that observes it. In other words, the fundamental constants and age of the universe are unremarkable, because human life and our minds are destined to arise in such a universe.

As previously stated, the FTA and anthropic principle may be explained by drawing upon some form of the multiverse theory, which states that there are universes other than our own. The argument follows: if there were an infinite number of universes with different physical laws and constants, it is unavoidable that some of them would have the exact combination of laws and constants that allow baryonic matter to evolve (coalescing into objects like stars, galaxies, and planets). Within those many universes, there would undoubtedly be some that were finely tuned to permit the existence of conscious beings like us. Thus, the multiverse theory makes our existence rather unsurprising; inevitable, in fact. An implication of this conclusion is that the conditions for our lives do not require an intelligent designer.

Current "theories of everything" support the multiverse by generating large numbers of universes with varying physical constants; therefore, it may be a plausible explanation for the FTA. However, the fact that there is no observational evidence that we live within a multiverse, combined with the fact that some of the theories are unfalsifiable, should make us reluctant to accept it as the answer.

The author Douglas Adams explains the anthropic principle with the following the analogy: "If you imagine a puddle waking up one morning and thinking, 'This is an interesting world I find myself in — an interesting hole I find myself in — fits me rather neatly, doesn't it? In fact it fits me staggeringly well, must have been made to have me in it!"

Infinite hexagons

In *The Library of Babel*, Jorge Luis Borges describes a library that holds books containing every possible combination of letters and numbers. The implications of such a library are hard to put into words. It would not only contain every essay, poem, book, and piece of scientific literature that has ever been written, but also every one that could ever be written.

Borges describes the Library as a "perhaps infinite" array of adjacent hexagonal rooms, with each room containing five bookshelves. Each bookshelf supports thirty-two 410-page books, with pages consisting of all possible combinations of letters and numbers. As a result, this library contains every phrase, sentence, word, etc. that could ever be written. Since we are discussing literally every possible arrangement of letters, the library would also contain an unfathomable number of unsensible books. For example, the following sentence is (by definition) included in the Library: "jkhjew2 r3jr3r2 hfjdsald3 43dsafgzzzzx." Also included in the Library is this sentence: "In the beginning God created the heaven and the Earth. And the earth was without form, and void....", and this one: "In the beginning God created liquor and cigarettes...." Of course, these examples could be changed by simply replacing the letter "a" with "b", and then "b" with "c", and so on.

The mathematics of Borges' Library puts its vastness in perspective. With ~1.3 million characters in each book and an alphabet of 25 letters, the total number of books in the library is 25 raised to the 1.3 millionth power (two followed by 1.8 million zeros). For each book in the Library, there are >30 million others that differ from it by only a single character. This number of books would fill the observable universe many more than a thousand trillion times.

The numbers are interesting, but the implications are more important. The Library contains the answer to every question we have. Every scientific and every philosophical question is answered from the origin to the meaning of life.

The Aleph

In a 1945 short story, Borges describes a point in space that contains all other points. This point in space, sharing the name of the story, is called *The Aleph*. When someone looks into the Aleph, they see everything in the universe from every angle simultaneously in an unbounded moment, with no superposition and no transparency. No overlap. Seemingly impossible, of course—but so is infinity itself.

The protagonist of Borges' story describes looking into the Aleph in the following manner:

"I saw a small iridescent sphere of almost unbearable brilliance. At first I thought it was revolving; then I realized that this movement was an illusion created by the dizzying world it bounded. The Aleph's diameter was probably two or three centimeters, but all space was there, actual and undiminished. Each thing (a mirror's face, let us say) was infinite things, since I distinctly saw it from every angle of the universe. I saw the teeming sea; I saw daybreak and nightfall; I saw the multitudes of America; I saw a silvery cobweb in the center of a black pyramid...I saw bunches of grapes, snow, tobacco, lodes of metal; steam; I saw convex equatorial deserts and each one of their grains of sand...I saw the oblique shadows of ferns on the floor of a greenhouse, saw tigers, pistons, tides, and armies, saw all the ants on Earth; I saw the coils and springs of love and death..."

The list of experiences goes on for an entire page.

I've always felt that the concept of infinity cannot be grasped with the arbitrary grunts and moans we call language. How can something timeless and dimensionless be described by time- and dimension-bound symbols? Borges' fiction has brought me closer to what I think infinity represents through feeling. Feeling is deep and wide—it doesn't seem restrained by physics (although it probably is). I think if I ever see what infinity is, it will be via experience, indescribable by language. And I won't ever get a hold on it like we humans are so desperate for, but instead will be flashed tiny glimpses, like sunlight through a pinhole.

Moore's Law

The first supercomputer that used a hard drive was the IBM 305 RAMAC—it weighed 1,000 kilograms, cost a 2023 equivalent of $30,000, and had a maximum storage capacity of 3.75MB. Today, around 60 years later, the baseline storage in a substantially cheaper 200-gram mobile phone is 64GB—a value 17,000 times greater than that of the behemoth described above. How did this happen?

In 1965, the engineer Gordon Moore observed that the number of transistors in an integrated circuit was doubling each year (ten years later, he adjusted this value to a doubling every two years and predicted the trend to continue indefinitely). For those unfamiliar with the guts of a computer, this essentially predicts that the processing power of computers will double every two years. This prediction of exponential growth has remained true in the decades since, and the result is a dramatic decrease in size and cost with a massive increase in computing power. To understand how this trend, known as *Moore's Law*, has transformed computation, it is necessary to have a working understanding of the transistor.

A transistor is an electrically driven switch that allows or denies the passage of a current through two terminals (known in modern computing as the source and the drain). Between the source and the drain is the "gate", which determines whether current can flow from one end to the other. Switching on and off this flow through transistors corresponds to the binary 1s and 0s that control digital computers. The reason that computing power has doubled each year is because of our ability to decrease the size of the transistor, which is represented by the distance between the source and the drain. In 2019, Samsung had a commercially available microprocessor with ˜2 trillion transistors at 7 nanometer node size. In 2020, several companies began mass producing transistors with a node size of 5 nanometers, which is approximately 0.005% the width of a human hair. In 2021, IBM unveiled new semiconductor chips with 2-nm transistors. The scale of these changes in computation simply blows my mind. Heroic feats of engineering that have transformed the history of civilization forever.

Many engineers have pointed out that Moore's law may be coming to an end, as the size of transistors cannot be made any smaller without quantum tunneling effects through the transistor gate. For our species to fulfill its potential, we will either need to overcome this size problem or reformulate modern computation.

Wheat and the chessboard

"Exponential is one of those technical words like 'momentum' or 'quantum' that originated in a scientific context with a precisely defined meaning but have entered the common jargon because they connote a useful concept that is not already adequately communicated in our everyday language." –Geoffrey West, *Scale*.

This sentence resonated with me and served as a reminder that the term *exponential* is seriously overused and misunderstood in modern society. To demonstrate the implications of exponential growth, it is helpful to call on what is known as the "wheat and chessboard problem." One version of the story goes as follows.

When Sessa—the ancient Indian Minister who is credited with inventing the game of chess—showed the king of Persia his creation, the king was so enthralled that he offered him any reward he desired. Sessa (much to the amusement and possible offense of the king) requested grains of rice, apportioned in the following manner: 1 grain on the first square of the chess board, 2 on the second, 4 on the third, 8 on the fourth, and so on, doubling the number of grains on each successive square. The king gave his treasurer the task of filling this request for Sessa and, at the end of the week, was told it would take more than all the kingdom's assets to complete the deal. In fact, the request was impossible to fill, and the reason why exposes the underappreciated magic of exponential growth.

It takes no more than a smart phone calculation to determine the number of rice grains Sessa was requesting. On the 64th and final square of the chessboard, the number of grains is equivalent to multiplying 2x2 63 times. Therefore, just to fill the 64th square would require 9,223,372,036,854,775,808 grains (just short of 10 million trillion grains—an amount that would form a pile the size of Mount Everest. The total number of grains on the board is approximately 18 million trillion.

1	2	4	8	16	32	64	128
256	512	1024	2048	4096	8192	$2 \cdot 10^4$	$3 \cdot 10^4$
$7 \cdot 10^4$	$1 \cdot 10^5$	....		....		....	
....		....		....	$2 \cdot 10^{18}$	$5 \cdot 10^{18}$	$9 \cdot 10^{18}$

Recursive self-improvement

Above we discussed what the physicist Geoffrey West calls "...the ultimate absurdity of exponential growth." That is, the profound and often misunderstood implications of exponential increase, whether it be counting grains of rice on a chessboard or the number of transistors that fit on a microchip. One theoretical exponential curve that has captured my attention is that representing the intellectual growth of an artificial agent capable of recursive self-improvement.

In his book *Superintelligence*, Nick Bostrom defines recursive self-improvement as follows: "The process in which an

AI iteratively improves its own intelligence—using its increasing intelligence to apply increasingly strong optimization power to the task of cognitive self-enhancement."

It is easy to visualize how this process would lead to exponential growth: an early AI designs a smarter version of itself, and then the improved version (let's say twice smarter) designs a smarter version of itself, and so on indefinitely. This kind of process is most likely to be responsible for what scientists and philosophers call an "intelligence explosion"—where a computational system goes from subhuman to radically superintelligent in a short period. This gets to the core of another fascinating characteristic of exponential growth: it starts slow but gets completely out of control once it is unleashed. In the context of AI, this is something to be concerned about. It predicts that all attempts to build an AI could be failures until a final critical component is put in place, at which point an AI would become capable of recursive improvement and blasting forth from its creators.

The "control problem" is another serious concern intimately linked with the concept of an exponentially improving machine. The idea is that it will be easy to create an AI that either misinterprets or disregards our initial commands. A common example is that of an AI designed to maximize the production of paper clips. Seemingly innocuous at the surface, those concerned point out that if not given far more specific constraints, it is plausible that an AI could use all the resources on our planet (and others) for the sole purpose of making paper clips. All this to say: if we are not careful to align an AI's goals with what maximizes the well-being of Homo sapiens, we may create our own destroyer. And with the abrupt and consistently unpredictable nature of exponential growth, it could happen at any moment.

AI as an existential risk

The more I read about it, the more convinced I am that the creation of an artificial intelligence should be recognized as a potentially devastating step in human evolution. Experts have emphasized that people do not consider AI a significant threat due to only a surficial knowledge of what an AI is. Due to the way that AI has been popularized and portrayed in science fiction, people asked to consider an AI commonly think of robots. In reality, the fields of AI and robotics are independent of one another. A majority of the current research in artificial intelligence does not consider robotics in the slightest, and when a general or superintelligent AI is created, it will likely not inhabit any human-like form. AI will likely not be limited by the mechanisms by which we humans have to navigate the physical world. A superintelligence would likely exploit the internet, vastly improve it (beyond our wildest imagination), and be able to infiltrate every nook and cranny of our digital infrastructure. In other words, I think that an AI will be capable of being everywhere at once, and there is no reason to think that we will be able to outsmart a superintelligence and simply shut it down if its goals do not align with ours. Lastly, the risk inherent in the creation of an AI is not necessarily that an AI will want to destroy Homo sapiens (as seen in *The Terminator*). More likely, we would be destroyed by an AI's indifference towards us as it fulfilled its goals. Consider the amount of thought we give to destroying ants as we build a new home or a Wal-Mart. It is plausible that our level of consciousness will be to an AI what an ants is to us.

Artificial speed

There are thousands of brilliant humans working tirelessly to create an artificial intelligence. Even more individuals are working on creating technology that will unintentionally assist in achieving this goal. As computing power doubles every two years (Moore's Law), we are getting closer every day to creating echoes of the human mind in the machine. As exciting as our technological growth is, we must maintain a proper level of concern regarding AI and consider the future paths we will need to take and strategies we will need to employ to control it.

In *Superintelligence*, Nick Bostrom outlines (in sometimes excruciating detail) the steps necessary to prevent AI from catastrophically getting away from us. Much of his book focuses on an artificial superintelligence—one that far exceeds a normal "general" artificial intelligence. Through recursive self-improvement (continuously building better versions of itself), it is possible that the creation of a general AI could inevitably lead to a machine that will leave us in the dust intellectually (and spiritually, artistically, etc). Superintelligence aside, I want to emphasize the possible implications of creating a computer matching only the intellectually capabilities of humans.

Let's say that in the year 2030, scientists and engineers create the first ever AI with intellect matching that of your average researcher at the University of Arizona or Caltech. Even with equivalent intelligence, a general AI would far exceed our capabilities through speed alone, as electronic circuits function approximately one million times faster than biochemical ones. Therefore, over the course of one human workday in the UA lab, the AI will have made ~20,000 years of intellectual progress. Imagining this occurring day after day gives an idea of what would become possible. We are talking about machines that will have the capability to read every book, every newspaper/magazine article, every legal document, and every webpage that has ever been written—in a matter of seconds. I am frightened by that alone.

Anatta

As far as we know, we are composed entirely of atoms. There is no evidence to suggest that there is a metaphysical self chillin' inside our bodies. I once found this conclusion of no soul unsettling and cold, but now view it in the opposite sense. The realization of the illusion of self is one that I am not sure I have come to fully understand, but my initial exploration of it has been exciting.

Some religions reassure the followers that there is something sacred about them as a person—that they each have a unique self or soul that will exist for all eternity. People are afraid of ceasing to exist, and that might be a reason why many are religious. I don't understand why existing eternally is longed for by our species. It seems to me that one of the truths of existence is that impermanence is what gives any object meaning. So why do we subscribe so quickly to the idea of an ever-lasting soul? If you have infinite time to experience after death, why would you even start doing anything if it could be put off for 20 trillion years with no consequence? To me, impermanence is what gives anything meaning—love, music, a flower, consciousness. Recognizing that my consciousness will most likely not transcend my body after death makes each moment so much richer and more meaningful. I enjoy exploring religious teachings but do not feel comfortable concluding that one of them is truer than another. One of the worst possible things we can do as a species is close our mind to possible explanations of the cosmos that don't align with what we believe.

Regardless of your beliefs, we are a collection of atoms studying atoms. It's one of the most ridiculous truths of the universe we occupy. All love.

Who are we talking to?

"Shit, I forgot my keys," echoed inside my dome as I left my apartment a few days ago. Internal dialogue is no uncommon thing, but I stopped in my tracks and was particularly weirded out by this one. I stopped and thought about the fact that I knew I had forgotten my keys, yet I reminded myself (?) via internal conversation. Plus, the natural subsequent thought was, "You need to go inside and find your keys." Who was I talking to? Who are all of us always talking to!? I think it is so interesting that we talk to ourselves as if we are separate from the things going on around us. I also feel it is hilarious how difficult it is to shake the feeling that the self exists and that when we relate to its existence, we cannot escape it! Haha, consider how easily you can remedy a roommate or partner that is annoying you—you can simply migrate elsewhere. We are given no such option when it comes to our own minds. We can become trapped in our own head with a self that is not there. I suppose meditation (whatever form that takes for each person; maybe skiing, maybe art) is analogous to migrating away from your irritating roommate.

We are not riding around inside of our bodies! There is no evidence to suggest that there is a separate entity (soul, self, etc.) residing somewhere between the ears and behind the eyes to whom all experiences are happening. It seems more likely that our overly complex human brains have deluded us into feeling that we primarily exist inside of our head. Realizing that this is not the case is liberating, exciting, and borderline magical. This feeling of separation also deprives us of the true beauty of existence. How much richer a life to be an integral part of the universe than a lonely soul detached from it!

Split brain

The corpus callosum (CC) is a thick nerve tract that facilitates neuronal communication between the two hemispheres of your brain. When you are acting in the world, information (such as sensory data) commonly enters the brain into one hemisphere, and is then linked to the other side via the approximately two hundred million nerve fibers (also known as "axons") of the CC. For example, when you pick up a wine glass with your left hand, sensory information is sent up to the right side of your brain (counterintuitive). The CC then carries that information over to the other side.

Mid-twentieth century surgical operations performed to reduce the frequency of epileptic seizures (excessive activity in the cortex, across the CC) exposed something profound about the CC and the human brain. They revealed that you can have your brain cut down the middle without dramatically altering your sense of self or the way you experience the world. But hidden behind this observation is an even more remarkable fact: split-brain patients appear to have two streams of consciousness within one skull.

This phenomenon of dual consciousness is backed by a condition known as the "alien hand syndrome", in which split-brain patients apparently stop themselves from going through with an action. It has been demonstrated and replicated (e.g., Ridley et al., 2016) that split-brain patients, when asked to pick up a glass, commonly interfere with their opposite hand—as if someone or something else had reacted differently to the instruction. This, and several other lines of evidence (see Sperry, 1982; Gazzaniga, 2005), point towards two separate consciousnesses existing within each person's brain (although it remains a topic of controversy in modern neuroscience). One of the pioneers of split-brain experiments, Roger Sperry, put it more eloquently: "...both the right and left hemispheres may be conscious simultaneously in different, even in mutually conflicting, mental experiences that run along in parallel" (Sperry, 1982).

The next time you are sitting in a quiet room by yourself, ponder the fact that you might not be alone...

The illusory self

The illusion that our consciousness is separate from our bodies and surroundings has permeated cultures across all human history. Some religious doctrines incorporated it into their belief systems, stating that the self or "soul" is an immortal entity that transcends biology and death. Other religions revolve around a recognition of the self as an illusion and have created a myriad of different paths to "enlightenment" or freedom from the self. Regardless, the illusion of the self has its claws sunk deeply into the trunk of human experience.

Few people better articulate the fundamentals of the illusory self for laypeople like me than the neuroscientist Sam Harris. In a popular YouTube video, he says the following (I am paraphrasing heavily):

The sense that we all have of riding around inside our own heads as a kind of passenger in vehicle of the body is an illusion. Most people don't feel identical to their bodies, they feel like they have bodies. They feel like they're inside the body...and most people feel like they are inside their heads. That sense of being a subject inside the head is an illusion. It makes no neuroanatomical sense...and there is no place inside the brain for your ego to be hiding."

Speaking about the neurophysiological processes that are likely responsible for consciousness, he states that, "Experience is a process...and there is not one unitary self that is carried through from one moment to the next." For some reason, that is the statement that always reignites my interest in the concept of self. That we are not carried through time/experience like a raft on a river. There is no passenger on the raft. The thought always leaves me unsettled. Harris goes on to emphasize it is possible to lose this feeling of self, or as he puts it, "...to have the center drop out of experience."

The point at which we shake the illusion of the self is the moment that we are able realize that we are not separate from our experience. We are not observers of the color, the light, the sound, the touch, or the cosmos. We are it.

The elements of the self

The illusion of self is best understood through a discussion of what it "feels like" to be a human. We spend most of our lives feeling that we are looking out at the world, as if we exist behind the eyes. Consequently, we feel that we are separate from our body. We are under the impression that our real self is simply hosted within our body. The most fundamental Buddhist teachings and many respected modern scientists recognize that this is illusory. Nothing suggests that there is someone behind experience to whom it is all happening.

In understanding selflessness, it needs to be made clear that the "self" is not something that is within you that you are trying to get rid of. From the beginning, the idea of self is illusory, and it has never in fact existed.

When we look at the clear night sky, the trained eye easily spots prominent constellations—like Orion, Ursa Major (Big Dipper), Leo, etc. These constellations don't really exist; what happens is we look at the sky and observe collections of stars that stand out and put a name on them as a single entity, separating them from other stars in the sky. The concept of self is like Orion. We recognize the elements of our existence (mind and body) are in a certain pattern, and then we subsequently assign a name to the elements (e.g., Jonathan). The crucial realization is that there is no self apart from the pattern, just as there is no Orion.

On a clear night, go outside and see if it is possible to not see Orion or the not see the Big Dipper—you will realize it is difficult. This is the same problem we face regarding our illusory self.

Bohr and the abyss

The physicist Niels Bohr, like many geniuses of the past, came close to losing his mind. An obsession with the interrelationships of nature, internal religious conflict, and deep contemplation on the self and infinity overwhelmed him (rightfully so). In his early 20's, Bohr articulated one of his concerns:

"[I start] to think about my own thoughts of the situation in which I find myself. I even think that I think of it, and divide myself into an infinite retrogressive sequence of 'I's' who consider each other. I do not know at which 'I' to stop as the actual, and the moment I stop at one, there is indeed again an 'I' which stops at it. I become confused and feel a dizziness as if I were looking down in a bottomless abyss."

I think most of us have considered and been overwhelmed by the abyss that Bohr describes. It doesn't take much meditation or reasoning to recognize that what we consider "I" or "the self" is an illusion. "I" is in a state of constant change; like a river, impossible to pin down. The river analogy is solid because changing variables are not solely limited to flow characteristics (like discharge), but also the shape of the channel itself. Just as a channelized flow constantly changes its own geometry, our present moment self shapes our surroundings. I don't think we are anything more than an "...infinite retrogressive sequence of 'I's..."

On having no head

"...what I found was khaki trouser legs terminating downwards in a pair of brown shoes, khaki sleeves terminating sideways in a pair of pink hands, and a khaki shirtfront terminating into absolutely nothing whatsoever!" This is how Douglas Harding recalls his experience of "...having no head." He realized that during a meditation with the eyes open, one does not experience their physical head. Rather, the head of the experiencer is replaced with the world. Or as he states, "I had lost a head and gained a world."

The claim of having no head may sound ridiculous at first, but contemplation and practice reveals that it is a profound realization, and one that gives insight into the nonduality of consciousness. In his book *Waking Up*, Sam Harris runs with this idea and instructs the reader to "...look for your head." So, try this—see if you can turn your attention in the direction of where you know your head to be. What do you see? It shouldn't take intense effort for you to experience, however briefly, the feeling of selflessness. What exists behind the eyes is not "you" or the "self", but rather experience itself. As stated above, the world exists where it feels as if you do. Harding explains this experience in a flowery fashion (my favorite):

"I fail to find...a viewer who is distinguishable from the view. Nothing whatsoever intervenes, not even that baffling and elusive obstacle called "distance": the huge blue sky, the pink-edged whiteness of the snows, the sparkling green of the grass—how can these be remote, when there's nothing to be remote from? The headless void refuses all definition and location."

Eternalism vs. presentism

The block universe theory of time (also known as *eternalism*) describes spacetime as an unchanging four-dimensional block in which the past, present, and future all exist simultaneously. This philosophy of time differs from that of the classical view that time is divided into three discrete regions (past, present, future), where the future is undefined, the present is moving into the future, and the past is immutably fixed. This classical view of time is integrated with "presentism", which argues that the present is the only moment that truly exists, and that the future and past are simply concepts used to describe the ever-changing present. When discussing the philosophy of time, it is easy to get tangled in linguistic weeds, so let us simplify the two arguments above in one statement: eternalism views past, future, and present existing concurrently; presentism states only the third exists.

I have a surficial knowledge of the ontological nature of time, but this idea of a block universe captured my attention. I am particularly interested in the support that eternalism gets from the Einstein's special and general theories of relativity. Broadly speaking, special relatively states that observers in different frames of reference record the same event happening at different times. Without going back into the details of general relativity, both theories support the argument that there is no physical basis for the present moment (i.e., occurrence can exist in time other than the present). I cannot restrain myself from mentioning determinism, or the idea that all events are ultimately determined by causes external to human free will. If we do indeed live in a block universe, all the choices we will make during our lives already exist.

von Neumann probes

In *Superintelligence*, Bostrom discusses methods that a sufficiently advanced artificial intelligence could use to take over the universe. One of the methods an AI could use is utilization of so-called von Neumann Probes. When considering possible methods of interstellar space exploration, the polymath John von Neumann theorized the idea of von Neumann probes, which are self-replicating spacecraft. A self-replicating spacecraft would have the ability to travel to a nearby celestial body (moon, star, asteroid, etc.) and use the raw material of the latter to create replicas of itself. The replicas would depart the body and each would travel to another, making more replicas of itself. This is a good example of exponential growth, as the growth rate would become more rapid in proportion to the growing total number of replicas. It is theorized that von Neumann probes, utilizing relatively conventional methods of space travel, could spread through that entire Milky Way galaxy in as little as 500,000 years.

von Neumann probes also have implications for our previous discussion on Fermi's paradox (where are the extraterrestrial visitors?). Some have argued that since we have not seen any self-replicating spacecraft, it is proof that ET intelligences do not exist. Carl Sagan subsequently pointed out that any sufficiently intelligent species would refuse to create an initial von Neumann probe, as it would inevitably lead to the consumption of the galaxy for spacecraft making purposes.

This brings us back to Bostrom's argument regarding artificial intelligence using self-replicating machines. If an AI was programed with a resource-based final goal, it is possible that the AI would utilize space travel to exploit all the resources in space. If the AI was not initially programmed to obtain only a certain amount of resources for its final goal, it is suggested that von Neumann probes would continue to replicate indefinitely, ultimately consuming the entire galaxy and universe, without even the slightest hesitation to destroy other civilizations or the one that created it.

The cosmic endowment

The *cosmic endowment* comprises the approximately 4×10^{20} stars that could potentially be reached by probes originating from Earth, before the expanding universe carries those stars over the cosmological horizon (Parfit, 1984; Bostrom, 2013). In essence, it consists of all the material/energy that our species has to work with (assuming we are alone in the universe). Assuming that we will determine how to colonize the observable universe (perhaps with the von Neumann probes described above), there are simple calculations that we can use to estimate the number of potential human lives in the future.

Parfit pointed out that if the Earth alone can remain habitable for a billion more years with a sustained population of >1 billion humans, there is potential for 10×10^{16} lives of normal duration (over a million times more lives than are currently being lived on Earth). In *Superintelligence*, Nick Bostrom takes this line of reasoning to the extreme, calculating both the number of biological and "cybernetic" human lives that could potentially be lived if we utilize the entire cosmic endowment. Cybernetic lives are digital implementations of lives (such as whole brain emulation) that could be possible in the future. First, Bostrom predicts that the cosmic endowment could support 10×10^{34} biological lives (if we assume that 10% of stars have a planet that is suitable for humans, as well as a billion-year long existence on each planet). For cybernetic human lives, we can convert the total energy that can realistically be harnessed from each star within the endowment to operations/sec (computational speed). This calculation predicts that at least 10×10^{58} 100-year human lives could be emulated if we are able to use the energy of all stars in the universe.

10,000,000,000,000,000,000,000,000,000,000,000,000, 000,000,000,000,000,000,000 potential human lives (which Bostrom argues is a gross underestimate) are hanging in the balance. "If we represent all the happiness experienced during one such life with a single teardrop of joy, then the happiness of these souls could fill and refill the Earth's oceans every second, and keep doing so for a hundred billion billion millennia." (Bostrom, 2014).

Celestial neighbors

This morning I was pondering what it would be like if Earth was the only celestial body in the Solar System. What if our planet's orbit was the only one that existed? No Venus, no Moon, no Mars, no Neptune, no Oort Cloud...imagine that! Aside from having a significant effect on gravitational dynamics and greatly limiting our knowledge on planetary formation and solar system evolution, I think it would have profound psychological effects as well. Such a great deal of our enthusiasm for space exploration is fueled by the eventual expeditions to the Moon and Mars; if these objects were taken away, what would we be striving to accomplish? Where would our ultimate exploration goals lie? Would we have any? Perhaps if our planet was the only thing between the Sun and our neighboring star we would take better care of it—or at least prioritize sustainability into the future.

But we are not alone. The Moon and Mars await us. Our generations will see the return to these celestial bodies, and future ones will likely have the privilege to settle them, explore them, and uncover their secrets. I feel optimistic about this.

The future of humanity

Bostrom (2013) developed four broad scenarios for the future of our species that I will address here. These potential futures include posthumanity, plateau, recurrent collapse, and extinction (see accompanying figure).

The most optimistic future would be posthumanity, consisting of the elimination of suffering, boundless population growth, and transcendence of biology (immortality). Also with posthumanity would likely come the expansion of our species to other planets, star-systems, and ultimately other galaxies. The plateau hypothesis implies that we are currently close to the pinnacle of our evolution and/or technological advancement. Some argue that this might occur because of the development of a stable political system—one that has ultimate control over the governed population. Another conceptual possibility is that soon our technological advances will be limited by the laws of nature. Maybe our computing power will plateau when the laws of nature do not allow us to make processors any smaller. The recurrent collapse scenario would consist of a pattern of collapse and regeneration. Possible causes of civilization collapse could be nuclear holocaust, climate change, starvation, etc. These are events that have the potential to but may not cause wholesale extinction. Rather they could cause near

THE FUTURE OF HUMANITY

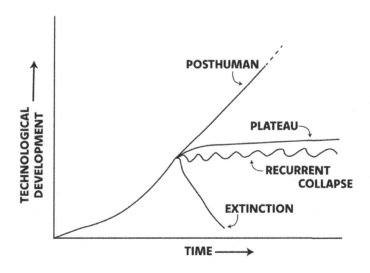

extinction and send us back hundreds or thousands of years intellectually. Subsequent regeneration of an inherently unstable civilization would then collapse in the same fashion. Finally, we arrive at extinction—the total disappearance of Homo sapiens from the universe. This could happen as a consequence of our evolving into a new species or life form, or simply dying out and non-continuation. Our species faces numerous existential threats that we hardly discuss. I highly recommend reading *The Precipice: Existential Risk and the Future of Humanity* by Toby Ord for a comprehensive and digestible overview of this topic.

A letter from utopia

In the spring of 1953 the president of the Pacific Telephone Co., Mark Sullivan, stated the following:

"Just what form the future telephone will take is, of course, pure speculation. Here is my prophecy: in its final development, the telephone will be carried about by the individual, perhaps as we carry a watch today. It probably will require no dial, and I think users will be able to see each other, if they want, as they talk. Who knows—it may actually translate from one language to another."

What was pure speculation has come to be. What was science fiction has become reality. Yet we seem not to realize this...we long for something else. Something more. What is it that we want? In times of global uncertainty, I wonder where we are going and why we even bother. But I refuse to let meaninglessness win. I don't think humans will die, at least not for a long time.

I like to think that words cannot convey the wonder of Homo sapiens future. That the most astounding well-being any of us have ever felt has become the norm. That our ability to mindfully engineer our genome and become one with our computational power has left our consciousness deep and wide.

That our lives are long, much longer. That we exist as we want, form and shape optional. That we can still love and appreciate the bliss of art and music and food and sex—that we experience these things far more powerfully than we once did. That all cultures have fused and are accepted, and each still has its own richness. That we are in control of an existence that surpasses our wildest imaginations. I hope these aspects of the future are only scratching the surface of its glory.

Immortal man

"Your body is a deathtrap. This vital machine and mortal vehicle, unless it jams first or crashes, is sure to rust anon. You are lucky to get seven decades of mobility; eight if you be Fortuna's darling. That is not sufficient to get started in a serious way, much less to complete the journey. Maturity of the soul takes longer. Why, even a tree-life takes longer!"

Nick Bostrom emphasizes the problem of our fragile and vulnerable human bodies. In this essay (Bostrom, 2010), he outlines three fundamental transformations that must be made if we wish to propel into the future towards Utopia. The first of these transformations is the essential change that must be made to increase (or indefinitely prolong) the human lifespan. In the quote above, he eloquently makes the point that for us to truly reach our potential would take far longer than our current bodies and minds allow. Our soft organs and brains break down and fall apart and we die before we can reach any significant height of intellectual maturity. He argues that our first step towards Utopia is to "...turn [our] biggest gun on aging, and fire." Taking control of our biochemical processes will allow us to continue to improve the system until we more or less become immortal. To quote Bostrom one last time, and emphasize the importance he places on human longevity, he states that "...any death prior to the heat death of the universe is premature if your life is good."

I've thought a fair amount about, if I had the option, whether I would choose to experience a dramatically prolonged or immortal existence. Initially, I considered the fact that it is the brevity of our existence that gives anything meaning. Like a flower blooming, it is knowing that it's beauty will only last a short period that gives it any beauty at all. However, after reconsidering and reading what Bostrom has to say, maybe prolonged life would ultimately be preferred, as unprecedented intellectual heights could be reached. Maybe experiencing these heights would outweigh the beauty and meaning found in our current short lives. It feels like at this point in my life I don't know the answer to anything haha – I guess I'll embrace that.

Immortal man, Pt. II

If you could extend your life, how long would you make it? 200 years? 500 years? 1,000? 1,000,000? The more I consider this question, the more content I become with the probable 75-80 years I will have on this planet. I may sound like a broken record on this point, but it seems self-evident that the meaning of human existence is only maintained by its brevity and fragility. I have been pondering this point in the context of human relationships—specifically, what would happen to our relationships if life were extended to great duration? The first conclusion I have come to is that it would be difficult to maintain an emotional attachment to other people if life was longer. I don't think it would be possible for relationships with others to remain fulfilling if we lived multiple centuries—more likely they would turn sour and we would like to detach. A second consideration is how we could bear the loss of someone we loved for multiple centuries or longer. I think that to avoid mourning our loved ones for centuries, we may choose to stand off and insulate ourselves from human attachment to begin with. Lastly, it is possible that we could outlive our fascination with life. Even with a life of 500 years (not to mention a million years), I could easily see Homo sapiens losing interest in the universe as a whole. We are driven to explore and discover because we are consciously aware that we don't have that long to do it!

Genetic engineering

In c. 65 AD, the Roman Stoic philosopher Seneca stated that "...[we] and the things we live amongst are destined to perish." In Joseph Goldstein's *The Experience of Insight*, I came across this concept rephrased as a reminder that "...inherent in all things that arise is decay." Although these are not original ideas of the two, I thought they served well as representatives of a concern we have had since the beginning of our self-consciousness.

The concept of impermanence is pervasive through all human history, reaching back as far as the earliest surviving work of human literature (the Epic of Gilgamesh). Philosophers, historians, and scientists have always written about death, decay, and mortality as if they were unavoidable universal truths, hence the examples above. Our views on mortality remained constant over thousands of years due to no scientific method and its advances. Only recently has this changed. The explosion of technology and science has profoundly transformed our understanding of life and death and we are beginning to question if mortality is indeed a "fundamental truth" of existence. In his book *Hacking Darwin*, Jamie Petzl addresses the current state of genetic engineering. His book makes clear that biology is no longer what it once was and will continue to change dramatically over the coming decades. Our technology and understanding of human biology are conjoining, and it is possible that within the next few decades we will have the ability to safely alter our genome and prolong our lives. Perhaps more importantly than prolonging our lives, we will be able to selectively choose which genes are present in our children, ultimately allowing us to produce offspring that exceed us in intelligence, strength, and creativity. If there are limits to this is something we will find out if we continue down this path. Maybe our period of genetic engineering will only play a minor role between where we are now and the eventual uploading of our consciousness to the Cloud.

I can't help but wonder what Seneca would think.

The Dyson Sphere

The advancement of our species has largely hinged on the ability to increase energy harvested over time. Early energy provided by the control of fire was followed by industrialization of our planet utilizing coal and oil/natural gas. Once we learned how to split an atom, atomic energy sourced from nuclear fission became possible (perhaps a form that has been underutilized), and we have since seen promising growth in form of renewable energy sources like solar and wind. In the next few decades, we may see another transition to energy from nuclear fusion, and assuming no planetary catastrophes, humanity will ultimately have complete control over Earth's resources. Once this point is reached, we would hopefully begin expanding outward into space. However, extensive space exploration and solar system colonization both require immense amounts of energy that cannot be realistically sourced from home base, so we must consider other sources. The ultimate solution to all energy needs would be the construction of what is known as a Dyson sphere.

Dyson spheres are hypothetical mega-engineering projects that would enable a species to capture a large proportion of their host star's energy output. These structures were originally conceived and popularized as solid spheres that completely envelop a star, but more realistic and recent designs involve series of large, orbiting panels ("Dyson swarm") that capture power from a star and beam it elsewhere. This would provide essentially unlimited energy for the species that constructs it, but it will require ungodly amounts of material. Calculations show that to create a swarm around our own sun would require thirty-quadrillion 1km x 1km panels, which sums to ~100 quintillion tons of raw material. Add on the energy required to assemble and transport that much material, we are looking at a daunting task indeed. Some scientists have suggested the complete disassembly of the planet Mercury for these purposes, as it resides relatively close to the sun, is metal-rich, has no atmosphere, and has a force of gravitational pull ~1/3 that of Earth.

Building a Dyson sphere is beyond our capabilities now and will be for a long time, but it is entertaining to consider what will be necessary for our species to become one with a widespread presence in the solar system. If you are interested in these ideas, I encourage you to check out Dyson (1960), the paper that popularized the concept of a Dyson sphere.

Technological singularity

The inventor and futurist Ray Kurtzweil claims that the singularity is near. Here, I refer not to the point of infinite density at the center of a black hole, but rather the *technological singularity*—a hypothetical point in time where advancement in technology becomes irreversible and uncontainable. In his thought-provoking 2005 book *The Singularity is Near*, Kurzweil outlines the exponential trends seen across almost every aspect of technology and introduces his law of accelerating returns, which predicts a continuation of exponential technologic growth until the singularity is reached. This runaway growth, he argues, will first result in humans transcending biology and becoming indistinguishable from the machine. Transcending biology could take several forms, but the most cited and most plausible (I think) would be something like uploading our consciousness to the cloud. No longer bound by the limitations of our weak and vulnerable bodies and minds, we would exist as pure intelligence, and the exponential growth would continue. After machine, man, computation, and intelligence all become indistinguishable, Kurzweil predicts an outward expansion of intelligence into the universe and eventually a complete intelligence saturation.

It is not easy to visualize a universe "...saturated with intelligence..." as Kurzweil puts it. Some futurists refer to this saturation as the expansion of a hypothesized form of matter known as *computronium*. Computronium is most easily understood as a rearrangement of existing matter that, when paired with super complex software, creates a substance that maximizes computation. Some theoretical work suggests that if the ordinary matter in a baseball-sized rock were converted to computronium, it would create a processor a trillion trillion times more powerful than all biological human brains today (Kurzweil, 2013). Kurzweil states that our destiny is to convert all matter in the universe to computronium, and he goes so far as to suggest that we will ultimately have control over the law of entropy and therefore the fate of the universe.

Laws of humanics

The human brain is arguably the most mysterious object in the known universe. Our science of consciousness is still largely limited to the philosophical and we have yet to develop analytical techniques that allow us to quantitatively investigate it. Although this science is young, there is no reason to think that we will not someday be able to gain an understanding of the mechanisms of consciousness. There is also no evidence to suggest that there are constituents of consciousness that escape the laws of physical reality (i.e., no suggestion of a metaphysical component of mind). Therefore, it may be possible in the future to mathematically define and simulate the human brain.

In Isaac Asimov's famous Robot Series, he formulates the *Three Laws of Robotics*, which are in essence mathematical descriptions of how the "positronic" (robot) brain functions. Asimov's novels leave the reader considering whether there are similar laws that govern the human brain. In theory, these laws could be expressed mathematically, allowing for broad predictions of the future. With our current science of the mind, we essentially must guess what is in store for us in the future and speculate what actions will make things better for humanity. A mathematical description of the mind would allow us to determine the right actions to take. If the brain can be quantified, it is interesting to ponder how many "Laws of Humanics" there are. One would imagine that the number of laws governing human action far exceeds the laws that would govern robots. Maybe the human brain is so complex that the number of laws would be infinite, and therefore impossible to define and simulate.

Lastly, let's say that the human brain could be perfectly defined and simulated. Would a perfect simulation of the human brain still be considered a simulation? What would separate the simulation from the real thing? If they're operating under identical initial conditions and subsequent laws, I don't think that they could be considered separate.

Evolving constants

Is it possible that the physical laws of the universe evolve over time?

In one of their broadcast conversations, Richard Feynman and cosmologist Fred Hoyle discuss how theoretical physics differs from other disciplines in natural science. Feynman outlines the fact that in most scientific disciplines there is a historical question, such as in geology and biology, where researchers study how the Earth's surface and life on it have evolved over time, respectively. That is, most scientists study what occurrences took place for things to become the way they are. Feynman points out that this is not the way that physicists approach their questions. They do not observe the physical laws operating in the universe today and ask, "How did they get that way?" or, "What occurrences led to the laws of nature as we see them?". Physicists think of the laws of nature as being static and unchanging. He points out that the physical laws of nature—ones that we commonly cite as being constant—may evolve over time, similar to the Earth's surface or the life that inhabits it.

Doesn't it seem like a big assumption that the laws of physics are constant? It seems like one of those almost dogmatic beliefs in science—that our existence is owed to unchanging physical laws. I don't think they would necessarily have to be constant for the universe to exist as we currently experience it. Most of the current theories and techniques that we use to explain the natural world rely on unchanging constants—think about how many of the equations you have used incorporate constants! What if these constants are just numbers that have dropped out of the math that we use to explain natural phenomena, and simply work because changes to the universal constants do not occur quickly enough to alter the answers?

Knot theory

Knot theory is the mathematical study of closed curves in three dimensions and how they can be translated/deformed without one part cutting through another. It helps to imagine taking a piece of string and looping, twisting, and biting it in various ways, and then fusing the loose ends together. One aspect of knot theory intends to describe how and if any given curve can be untangled (back into a circle for example). Another question (that is a real mind fu*%) is whether any two given curves represent different knots, or if they are actually the same knot that can be deformed into the other.

Knots can range in complexity, usually described by the number of times that a knot's projected shadow has crossings that go over one another. For example, the "unknot" is a simple circle with no crossings, and the simplest of all knots is the trefoil knot (AKA overhand knot), which has three crossings. The number of crossings in a knot allows for classification by "order"; a trefoil knot is a third order knot. As you increase knot order, the number of possible knots increases rapidly. This is nicely demonstrated by the fact that there are over a million known 16-order knots. The knots shown here are a miniscule proportion of the over 6 billion knots that have been documented since the birth of the subdiscipline in the early 1800's.

It is important and often overlooked that the scientific method can pop out unintended results that eventually play a huge role in the progress of humanity. Knot theory is a great example of something that may have seemed like a ridiculous use of time and resources in the 19th century and early 20th century—and then Watson and Crick determined the morphology of DNA. DNA consists of two polynucleotide strands twisted around each other in a double helix, and these strands are packed together into genes....knotted and coiled actually. Mathematicians are now using topological principles of knot theory to understand how DNA unknots itself, how it replicates and is transcribed, and ultimately how molecular biology operates.

It serves as a solid reminder that we need to fund science and technology for more reasons than only the intended goals and hypotheses. The accidents of science are invaluable.

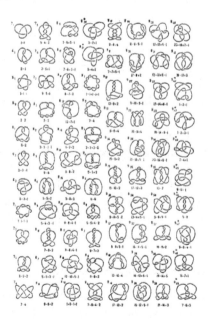

Some examples of knots. If you're interested in knot theory, see Burde et al. (2013).

Life before technology

Do you remember what it was like to live before computers, cell phones, and/or the internet? Isn't it amazing that our overlapping generations will be the last ones to ever know what it was like?

I, presumably like many of you, spent my childhood in a world that did not contain ubiquitous cell phones. And perhaps sadly, I only have vague, fading memories of life before it was inundated by technology. I had a conversation with some family about their experience living through the inflection point of an exponential change in technology. We reminisced on the times before search engines when it was necessary for one to search through the Encyclopedia Britannica to get an answer to nagging questions. My dad told me in a story that I found hilarious about the first laptop he bought in 1986—an IBM PC Convertible with a 16MB hard drive and 256kB RAM for $2,000. It was entertaining that people with deeper experience than I still had vivid memories of the times before screens. I was mostly curious if the people who had reached full adulthood before the technological explosion had had their memories jaded by it.

The technological changes that we have experienced and are currently going through are crazy. I don't know if it is a common feeling among each successive generation, but I feel that ours is an exceptional one (the one or two that lived through the nose of the flood). With all modesty, I think that the technological change we have seen in the last 20 years and will see in the next 20 could be the most consequential development in the history of human evolution. What we do with our tech and where we go from here is up to us.

The vulnerable world hypothesis

I am becoming increasingly interested in the philosophical domain that confronts what are known as existential threats. Nick Bostrom has popularized the idea that in the future, we will face risks that have the potential to cause complete annihilation of Homo sapiens. In a paper titled *The Vulnerable World Hypothesis* (Bostrom, 2019), a situation is outlined in which an existential risk could manifest out of the desire we humans have to invent new things. To better understand this hypothesis, he has developed a helpful thought experiment that we will call the *urn metaphor.* In this metaphor, Bostrom has the reader consider human creativity and invention as the process of pulling random balls out of the urn of invention. In this urn, there are three different types of balls—white, grey, and black. The white balls represent inventions that have been entirely beneficial for our species, and we have been fortunate to extract mostly this type since we started reaching in. The grey balls have both positive and harmful effects, like our ability to create nuclear fission (it can provide us with great amounts of energy, but it also has the potential to kill millions of people). Lastly, we come to the black balls. These are representative of inventions that "invariably or by default destroy the civilization that invents it." Bostrom lays out a strong argument not only for the existence of black-ball inventions, but also the likelihood of extracting one from the urn. If we continue making strides in science and technology, it seems inevitable that we will pull out a black ball, and once we have removed it from the urn, there is no putting it back. We cannot uninvent this potentially devastating technology (or whatever it may be).

The projection and the mold

"I believe forms and ideas exist distinct from matter and energy. For example, the form of a triangle with specified dimensions exists whether matter represents it or not, and it will continue to exist forever." –Josh Plovanic.

I have been confused by the fact that in an exchange of ideas, nothing physical ever becomes mobile; confused by what exactly is transported to the recipient of an idea. When I posed these questions to my friend quoted above, he responded with the following: forms (things like shapes and ideas) cannot exist in spacetime without a mold or cast that brings them into reality. He argues that all forms, whether as simple as a triangle or as complex as a braided river, exist independently of the physical substance that represents them (the mold). He used acting as a clarifying example, stating that actors portray their characters into our reality from a fictional or metaphorical reality that exists elsewhere. The actors represent the mold, while the characters they play are forms extracted from an independently existing reality. It is the case that forms and ideas exist eternally elsewhere and cannot manifest in our world without a representative. Plato suggested forms and ideas exist in a separate reality from matter and energy, are eternal, and that every idea that will someday manifest itself in a sense already exists as a blueprint.

Memes

I am a meme lord. Here, I'm talking about memes in the context of the internet. Poorly understood (at least by my generation) is that the actual term "meme" goes far beyond the internet and has philosophical significance.

The word 'meme' was coined by the famous biologist Richard Dawkins in his first book, *The Selfish Gene*. Derived from the Ancient Greek work '*mimeme*', meaning "imitated thing", Dawkins proposed the word to conceptualize the spread of cultural phenomena and ideas in the context of evolutionary theory. The actual definition provided by Dawkins states that memes are, "...ideas, behaviors, or styles that spread by means of imitation from person to person within a culture." His argument is that memes are cultural analogues to genes, having the ability to respond to selective pressure, mutate, and self-replicate. Indeed, it seems that ideas, styles, etc. react to pressures of competition and that they can undergo variation and mutation via their hosts. A simple example that makes clear the analogy with evolution by natural selection is that memes that propagate prolifically are less likely to become extinct, while others die out. The discussion surrounding memes is deep and entertaining, not to mention important. Dawkins points out that it is perhaps more likely to have good memes persist in nature than genes. Socrates is given as an example; a man who probably has none of his genes active on Earth, but whose ideas are pervasive.

All philosophy aside, I love internet memes. My sister and I have discussed the ability that they have to bring people together (especially nihilists haha) and recognize shared problems. They also often make harmless fun of people who probably deserve it (like astrologers).

Searching for a key

"It is a popular delusion that the scientific enquirer is under an obligation not to go beyond the generalization of observed facts...but anyone who is practically acquainted with scientific work is aware that those who refuse to go beyond the facts, rarely get as far." -T.H. Huxley, 1870.

This beautiful quote from Huxley reminds me of the story of a man who under a single streetlamp is frantically searching for something. When a passerby asks, "What are you looking for?", the searching man tells him that he has lost his house key. After searching together for several minutes, the passerby says, "Are you sure you lost it here?". The man replies, "Not at all, but unless I lost it here I will never find it!". Huxley's emphasis on looking beyond facts during scientific investigation is analogous to creeping beyond the scope of the streetlamp. It is there, he argues, that the far-reaching and impactful discoveries will be made. But how can we as scientists have truths revealed to us that are enveloped in darkness? I suppose the scientific method serves to gradually increase the amount of light, and what we are looking for slowly comes into view (even if it is still difficult to see). I must add that it seems occasionally that individual scientists come into this world who are not limited by the scope of the light (Newton, von Neumann, Einstein, etc.). They walk seemingly without direction into the dark and come back with the master key. God only knows how they do this[*].

One of the reasons I love the metaphor above is because of how accurately I think it represents the vast amount that is unknown to us. The light from a single streetlamp is so incredibly miniscule relative to the world that surrounds it, and I think that our current scope of and understanding is similar. Who knows what kinds of secrets exist in the murky waters or dark forests far separated from the streetlamp!

Have fun with it

I find it comical that I am nestled in between what is probably infinite time in both directions, yet I care about how people respond to the work I create. I am some side-effect of the universe, a vulnerable and improbable creature that has naturally evolved in a deterministic universe (triggered), yet I care about how people perceive me. Black holes—ravenous for matter—twist space and time into a knot and high-mass stars collapse and explode throughout the cosmos—meanwhile, I give attention to vapid Instagram stories. My consciousness (whether it is separate from my surroundings or not) is flying through space on a 6-sextillion kilogram sphere of rock and water, yet I ruminate on the past and have angst about the future.

Something that is perhaps stranger than our existence is how we use it. Hopefully it goes without saying that mindless social media browsing is a bad use of your unlikely existence. Sure, there is meaning to be found in challenging work, but the

---

[*]Hans Bethe has an amazing quote when describing John von Neumann, saying, "I have sometimes wondered whether a brain like von Neumann's does not indicate a species superior to that of man", and later he wrote that "[von Neumann's] brain indicated a new species, an evolution beyond man".

fact that we cling to results and care about analytics and recognition is strange. I like to remind myself of the things above as much as I can (of course, the list of existential absurdities goes on and on); it reminds me to not take things too seriously. That is a damn shame, you know, when people spend their entire lives being serious. To each their own, I suppose, but I don't vibe with it. Go outside and roll in the grass. Do a fu**ing somersault or something, you sad bastard. Look up at space (the top is down on convertible Earth). Be humbled. Go drink some beer or smoke a joint or something and laugh about the fact that you are a stack of atoms.

With all due respect, your existence is ridiculous. Probably meaningless in the long term. But you have experience right now—and that is just silly...

References

Adams, J. B., Gillespie, A. R., Jackson, M. P. A., Montgomery, D. R., Dooley, T. P., Combe, J.-P., & Schreiber, B. C. (2009). Salt tectonics and collapse of Hebes Chasma, Valles Marineris, Mars. Geology, 37(8), 691–694. doi:10.1130/G30024A.1

Allen, J. R. L. (1985). Principles of Physical Sedimentology. Springer Netherlands. doi:10.1007/978-94-010-9683-6

Bacon, F. (2019). Novum organum = New instrument. Anodos Books.

Becker, H. N., Alexander, J. W., Atreya, S. K., Bolton, S. J., Brennan, M. J., Brown, S. T., Guillaume, A., Guillot, T., Ingersoll, A. P., Levin, S. M., Lunine, J. I., Aglyamov, Y. S., & Steffes, P. G. (2020). Small lightning flashes from shallow electrical storms on Jupiter. Nature, 584(7819), 55–58. doi:10.1038/s41586-020-2532-1

Borges, J. L., & Hurley, A. (1998). Collected fictions. Penguin Books.

Bostrom, N. (2008). Why I Hope the Search for Extraterrestrial Life Finds Nothing. MIT Technology Review, May/June, 72–77.

Bostrom, N. (2003). Are We Living in a Computer Simulation? The Philosophical Quarterly, 53(211), 243–255. doi:10.1111/1467-9213.00309

Bostrom, N. (2010). Letter from Utopia. Studies in Ethics, Law, and Technology, 2(1). doi:10.2202/1941-6008.1025

Bostrom, N. (2013). Existential Risk Prevention as Global Priority: Existential Risk Prevention as Global Priority. Global Policy, 4(1), 15–31. doi:10.1111/1758-5899.12002

Bostrom, N. (2016). Superintelligence: Paths, dangers, strategies. Oxford University Press.

Bostrom, N. (2019). The Vulnerable World Hypothesis. Global Policy, 10(4), 455–476. doi:10.1111/1758-5899.12718

BP. (2022). BP Statistical Review of World Energy 2022 (71st edition; pp. 1–60).

Brakenridge, G. R. (2021). Solar system exposure to supernova γ radiation. International Journal of Astrobiology, 20(1), 48–61. doi:10.1017/S1473550420000348

Brown, S., Janssen, M., Adumitroaie, V., Atreya, S., Bolton, S., Gulkis, S., Ingersoll, A., Levin, S., Li, C., Li, L., Lunine, J., Misra, S., Orton, G., Steffes, P., Tabataba-Vakili, F., Kolmašová, I., Imai, M., Santolík, O., Kurth, W., ... Connerney, J. (2018). Prevalent lightning sferics at 600 megahertz near Jupiter's poles. Nature, 558(7708), 87–90. doi:10.1038/s41586-018-0156-5

Burde, G., Zieschang, H. & Heusener, M. (2013). Knots. Berlin, Boston: De Gruyter. doi:10.1515/9783110270785

Busse, F. H. (1976). A simple model of convection in the Jovian atmosphere. Icarus, 29(2), 255–260. doi:10.1016/0019-1035(76)90053-1

Carr, M. H. (1974). Tectonism and volcanism of the Tharsis Region of Mars. Journal of Geophysical Research, 79(26), 3943–3949. doi:10.1029/JB079i026p03943

Carrapa, B., DeCelles, P. G., & Romero, M. (2019). Early Inception of the Laramide Orogeny in Southwestern Montana and Northern Wyoming: Implications for Models of Flat-Slab Subduction. Journal of Geophysical Research: Solid Earth, 124(2), 2102–2123. doi:10.1029/2018JB016888

Carroll, S. (2007). Dark matter, dark energy: The dark side of the universe. Teaching Co.

Caylor, E., Carrapa, B., Jepson, G., Sherpa, T. Z., & DeCelles, P. G. (2023). The rise and fall of Laramide topography and the sediment evacuation from Wyoming. Geophysical Research Letters, 50(14), e2023GL103218.

Chalmers, D. J. (2010). The Character of Consciousness. Oxford University Press. doi:10.1093/acprof:oso/9780195311105.001.0001

Chiang, T. (2002). Stories of your life and others (1st ed). Tor.

Coney, P. J., & Harms, T. A. (1984). Cordilleran metamorphic core complexes: Cenozoic extensional relics of Mesozoic compression. Geology, 12(9), 550–554. doi:10.1130/0091-7613(1984)12<550:CMCCCE>2.0.CO;2

Coney, P. J., & Reynolds, S. J. (1977). Cordilleran Benioff zones. Nature, 270(5636), 403–406. doi:10.1038/270403a0

Davies, P. C. W. (2003). How to build a time machine. Penguin Group.

Dawkins, R. (2016). The Selfish Gene (4th Edition). Oxford University Press.

DeCelles, P. G. (2004). Late Jurassic to Eocene evolution of the Cordilleran thrust belt and foreland basin system, western U.S.A. American Journal of Science, 304(2), 105–168. doi:10.2475/ajs.304.2.105

DeCelles, P. G., Carrapa, B., Horton, B. K., McNabb, J., Gehrels, G. E., & Boyd, J. (2015). The Miocene Arizaro Basin, central Andean hinterland: Response to partial lithosphere removal? In P. G. DeCelles, M. N. Ducea, B. Carrapa, & P. A. Kapp, Geodynamics of a Cordilleran Orogenic System: The Central Andes of Argentina and Northern Chile. Geological Society of America. doi:10.1130/2015.1212(18)

Dyson, F. J. (1960). Search for Artificial Stellar Sources of Infrared Radiation. Science, 131(3414), 1667–1668. doi:10.1126/science.131.3414.1667

Emlen, D. J. (2014). Animal weapons: The evolution of battle (First edition). Henry Holt and Company.

Feynman, R. P., & Davies, P. (2011). Six easy pieces: Essentials of physics explained by its most brilliant teacher. Basic Books, a member of the Perseus Books group.

Fields, C., Hoffman, D. D., Prakash, C., & Singh, M. (2018). Conscious agent networks: Formal analysis and application to cognition. Cognitive Systems Research, 47, 186–213. doi:10.1016/j.cogsys.2017.10.003

Freeman, K. C. (1970). On the Disks of Spiral and S0 Galaxies. The Astrophysical Journal, 160, 811. doi:10.1086/150474

Gazzaniga, M. S. (2005). Forty-five years of split-brain research and still going strong. Nature Reviews Neuroscience, 6(8), 653–659. doi:10.1038/nrn1723

Goldstein, J. (1983). The experience of insight: A simple and direct guide to Buddhist meditation (1st Shambhala ed). Shambhala ; Distributed in the U.S. by Random House.

Gould, S. J. (1990). Wonderful life: The Burgess Shale and the nature of history. Norton & Co.

Gould, S. J. (2011). Rocks of Ages: Science and Religion in the Fullness of Life. Random House US. https://public.ebookcentral.proquest.com/choice/publicfullrecord.as px?p=5336686

Gradstein, F. M. (2012). Introduction. In The Geologic Time Scale (pp. 1–29). Elsevier. doi:10.1016/B978-0-444-59425-9.00001-9

Graham, M. J., Ford, K. E. S., McKernan, B., Ross, N. P., Stern, D., Burdge, K., Coughlin, M., Djorgovski, S. G., Drake, A. J., Duev, D., Kasliwal, M., Mahabal, A. A., van Velzen, S., Belecki, J., Bellm, E. C., Burruss, R., Cenko, S. B., Cunningham, V., Helou, G., ... Soumagnac, M. T. (2020). Candidate Electromagnetic Counterpart to the Binary Black Hole Merger Gravitational-Wave Event S190521g. Physical Review Letters, 124(25), 251102. doi:10.1103/PhysRevLett.124.251102

Greenberg, R. (2010). Transport Rates of Radiolytic Substances into Europa's Ocean: Implications for the Potential Origin and Maintenance of Life. Astrobiology, 10(3), 275–283. doi:10.1089/ast.2009.0386

Grinnell, F. (1996). Ambiguity in the Practice of Science. Science, 272(5260), 333–333. doi:10.1126/science.272.5260.333

Guillot, T., Li, C., Bolton, S. J., Brown, S. T., Ingersoll, A. P., Janssen, M. A., Levin, S. M., Lunine, J. I., Orton, G. S., Steffes, P. G., & Stevenson, D. J. (2020). Storms and the Depletion of Ammonia in Jupiter: II.

Explaining the Juno Observations. Journal of Geophysical Research: Planets, 125(8). doi:10.1029/2020JE006404

Hameroff, S., & Penrose, R. (2014). Consciousness in the universe: A review of the 'Orch OR' theory. Physics of Life Reviews, 11(1), 39–78. doi:10.1016/j.plrev.2013.08.002

Hardy, T., & Dolin, T. (2003). Tess of the D'Urbervilles. Penguin Books.

Harris, A. (2019). Conscious: A brief guide to the fundamental mystery of the mind (First edition). Harper Collins Publishers.

Harris, S. (2006). Letter to a Christian nation (1st ed). Knopf.

Harris, S. (2011). Moral landscape: How science can determine human values (1st Free Press paperback ed). Free Press.

Harris, S. (2014). Waking up: A guide to spirituality without religion (First Simon&Schuster hardcover edition). Simon & Schuster.

Hawking, S. (1998). A brief history of time (Updated and expanded tenth anniversary ed). Bantam Books.

Hawking, S. W. (1974). Black hole explosions? Nature, 248(5443), 30–31. doi:10.1038/248030a0

Hoffman, D. D., & Prakash, C. (2014). Objects of consciousness. Frontiers in Psychology, 5. doi:10.3389/fpsyg.2014.00577

Howlett, C. J., Reynolds, A. N., & Laskowski, A. K. (2021). Magmatism and Extension in the Anaconda Metamorphic Core Complex of Western Montana and Relation to Regional Tectonics. Tectonics, 40(9). doi:10.1029/2020TC006431

Iess, L., Militzer, B., Kaspi, Y., Nicholson, P., Durante, D., Racioppa, P., Anabtawi, A., Galanti, E., Hubbard, W., Mariani, M. J., Tortora, P., Wahl, S., & Zannoni, M. (2019). Measurement and implications of Saturn's gravity field and ring mass. Science, 364(6445), eaat2965. doi:10.1126/science.aat2965

Jackson, F. (1982). Epiphenomenal Qualia. The Philosophical Quarterly (1950-), 32(127), 127–136. JSTOR. doi:10.2307/2960077

Jacob, F., & Philip, F. (1995). The statue within: An autobiography. Cold Spring Harbor Laboratory Press.

Jacobson, C. E., Grove, M., Pedrick, J. N., Barth, A. P., Marsaglia, K. M., Gehrels, G. E., & Nourse, J. A. (2011). Late Cretaceous-early Cenozoic tectonic evolution of the southern California margin inferred from provenance of trench and forearc sediments. Geological Society of America Bulletin, 123(3-4), 485–506. doi:10.1130/B30238.1

Kapp, P., & DeCelles, P. G. (2019). Mesozoic–Cenozoic geological evolution of the Himalayan-Tibetan orogen and working tectonic hypotheses.

American Journal of Science, 319(3), 159-254. doi:10.2475/03.2019.01

Kurtzweil, R. (2013). What Will Happen After The Technological Singularity? https://www.youtube.com/watch?v=lAJkDrBCA6k

Kurzweil, R. (2005). The singularity is near: When humans transcend biology. Viking.

Lewis, S. L., & Maslin, M. A. (2015). Defining the Anthropocene. Nature, 519(7542), 171-180. doi:10.1038/nature14258

Li, L., MaBouDi, H., Egertová, M., Elphick, M. R., Chittka, L., & Perry, C. J. (2017). A possible structural correlate of learning performance on a colour discrimination task in the brain of the bumblebee. Proceedings of the Royal Society B: Biological Sciences, 284(1864), 20171323. doi:10.1098/rspb.2017.1323

Locey, K. J., & Lennon, J. T. (2016). Scaling laws predict global microbial diversity. Proceedings of the National Academy of Sciences, 113(21), 5970-5975. doi:10.1073/pnas.1521291113

Marcus Aurelius. (2014). Meditations.

McKenzie, D., & Nimmo, F. (1999). The generation of martian floods by the melting of ground ice above dykes. Nature, 397(6716), 231-233. doi:10.1038/16649

McMillan, M., & Schoenbohm, L. M. (2023). Diverse Styles of Lithospheric Dripping: Synthesizing Gravitational Instability Models, Continental Tectonics, and Geologic Observations. Geochemistry, Geophysics, Geosystems, 24(2), e2022GC010488. doi:10.1029/2022GC010488

Méndez Harper, J. S., McDonald, G. D., Dufek, J., Malaska, M. J., Burr, D. M., Hayes, A. G., McAdams, J., & Wray, J. J. (2017). Electrification of sand on Titan and its influence on sediment transport. Nature Geoscience, 10(4), 260-265. doi:10.1038/ngeo2921

Metzl, J. F. (2020). Hacking Darwin: Genetic engineering and the future of humanity.

Molnar, P., & Houseman, G. A. (2004). The effects of buoyant crust on the gravitational instability of thickened mantle lithosphere at zones of intracontinental convergence: Crust and gravitational instability of thickened mantle lithosphere. Geophysical Journal International, 158(3), 1134-1150. doi:10.1111/j.1365-246X.2004.02312.x

Moura, R. L., Amado-Filho, G. M., Moraes, F. C., Brasileiro, P. S., Salomon, P. S., Mahiques, M. M., Bastos, A. C., Almeida, M. G., Silva, J. M., Araujo, B. F., Brito, F. P., Rangel, T. P., Oliveira, B. C. V., Bahia, R. G., Paranhos, R. P., Dias, R. J. S., Siegle, E., Figueiredo, A. G., Pereira, R. C., ... Thompson, F. L. (2016). An extensive reef system at the Amazon River mouth. Science Advances, 2(4), e1501252. doi:10.1126/sciadv.1501252

Nagel, T. (1974). What Is It Like to Be a Bat? The Philosophical Review, 83(4), 435. doi:10.2307/2183914

Nagel, T. (1980). Mortal Questions. The Philosophical Review, 89(3), 473. doi:10.2307/2184400

Oort, J. H. (1940). Some Problems Concerning the Structure and Dynamics of the Galactic System and the Elliptical Nebulae NGC 3115 and 4494. The Astrophysical Journal, 91, 273. doi:10.1086/144167

Ord, T. (2020). The precipice: Existential risk and the future of humanity. Hachette Books.

Parfit, D. (1987). Reasons and persons (1. issued in paperback (with corr.), reprinted with further corr). Clarendon Press.

Penrose, R. (2016). The emperor's new mind: Concerning computers, minds and the laws of physics (Revised impression as Oxford landmark science). Oxford University Press.

Petri, G., Expert, P., Turkheimer, F., Carhart-Harris, R., Nutt, D., Hellyer, P. J., & Vaccarino, F. (2014). Homological scaffolds of brain functional networks. Journal of The Royal Society Interface, 11(101), 20140873. doi:10.1098/rsif.2014.0873

Pirsig, R. M. (Ed.). (2006). Zen and the art of motorcycle maintenance: An inquiry into values (1. HarperTorch paperback printing). HarperTorch.

Pollan, M. (2018). How to change your mind: What the new science of psychedelics teaches us about consciousness, dying, addiction, depression, and transcendence. Penguin Press.

Rampino, M. R. (2015). Disc dark matter in the Galaxy and potential cycles of extraterrestrial impacts, mass extinctions and geological events. Monthly Notices of the Royal Astronomical Society, 448(2), 1816–1820. doi:10.1093/mnras/stu2708

Rampino, M. R., & Stothers, R. B. (1984). Terrestrial mass extinctions, cometary impacts and the Sun's motion perpendicular to the galactic plane. Nature, 308(5961), 709–712. doi:10.1038/308709a0

Raynaud, R., Guilet, J., Janka, H.-T., & Gastine, T. (2020). Magnetar formation through a convective dynamo in protoneutron stars. Science Advances, 6(11), eaay2732. doi:10.1126/sciadv.aay2732

Ridley, B., Beltramone, M., Wirsich, J., Le Troter, A., Tramoni, E., Aubert, S., Achard, S., Ranjeva, J.-P., Guye, M., & Felician, O. (2016). Alien Hand, Restless Brain: Salience Network and Interhemispheric Connectivity Disruption Parallel Emergence and Extinction of

Diagonistic Dyspraxia. Frontiers in Human Neuroscience, 10. doi:10.3389/fnhum.2016.00307

Ronemus, C. B., Orme, D. A., Guenthner, W. R., Cox, S. E., & Kussmaul, C. A. L. (2023). Orogens of Big Sky Country: Reconstructing the Deep-Time Tectonothermal History of the Beartooth Mountains, Montana and Wyoming, USA. Tectonics, 42(1), e2022TC007541. doi:10.1029/2022TC007541

Rosling, H., Rosling, O., & Rönnlund, A. R. (2018). Factfulness: Ten reasons we're wrong about the world–and why things are better than you think (First edition). Flatiron Books.

Schwägerl, C. (2023). A Golden Spike Would Mark the Earth's Next Epoch: But Where? Yale Environment 360. https://e360.yale.edu/features/anthropocene-site-competition-golden-spike

Seneca, L. A., & Campbell, R. (1969). Letters from a Stoic: Epistulae morales ad Lucilium. Penguin.

Seneca, L. A., Reinhardt, T., & Davie, J. (2008). Dialogues and essays (Reissued). Oxford Univ. Press.

Simard, S. W., & Durall, D. M. (2004). Mycorrhizal networks: A review of their extent, function, and importance. Canadian Journal of Botany, 82(8), 1140–1165. doi:10.1139/b04-116

Simon, C. (2018). Can quantum physics help solve the hard problem of consciousness? A hypothesis based on entangled spins and photons. doi:10.48550/ARXIV.1809.03490

Sperry, R. (1982). Some Effects of Disconnecting the Cerebral Hemispheres. Science, 217(4566), 1223–1226. doi:10.1126/science.7112125

Stølum, H.-H. (1996). River Meandering as a Self-Organization Process. 271, 4.

Suomi, V. E., Limaye, S. S., & Johnson, D. R. (1991). High Winds of Neptune: A Possible Mechanism. Science, 251(4996), 929–932. doi:10.1126/science.251.4996.929

Sweetlove, L. (2011). Number of species on Earth tagged at 8.7 million. Nature, news.2011.498. doi:10.1038/news.2011.498

Taylor, J. (1975). Black Holes. Fontana/Collins.

Thaddeus, P., & Chanan, G. A. (1985). Cometary impacts, molecular clouds, and the motion of the Sun perpendicular to the galactic plane. Nature, 314(6006), 73–75. doi:10.1038/314073a0

The LIGO Scientific Collaboration, the Virgo Collaboration, Abbott, R., Abbott, T. D., Abraham, S., Acernese, F., Ackley, K., Adams, A., Adams, C., Adhikari, R. X., Adya, V. B., Affeldt, C., Agathos, M., Agatsuma, K., Aggarwal, N., Aguiar, O. D., Aiello, L., Ain, A.,

Ajith, P., ... Weltevrede, P. (2020). Gravitational-wave constraints on the equatorial ellipticity of millisecond pulsars. The Astrophysical Journal Letters, 902(1), L21. doi:10.3847/2041-8213/abb655

Thorne, K. S. (1994). Black holes and time warps: Einstein's outrageous legacy. W.W. Norton.

translated by Daw Mya Tin ; edited by the Editorial Committee. (1990). The Dhammapada: Verses & stories. First reprint edition. Delhi, India : Sri Satguru Publications, 1990.
https://search.library.wisc.edu/catalog/999668923302121

Uggerhøj, U. I., Mikkelsen, R. E., & Faye, J. (2016). The young centre of the Earth. European Journal of Physics, 37(3), 035602.
doi:10.1088/0143-0807/37/3/035602

Vasavada, A. R., & Showman, A. P. (2005). Jovian atmospheric dynamics: An update after Galileo and Cassini. Reports on Progress in Physics, 68(8), 1935–1996. doi:10.1088/0034-4885/68/8/R06

Vine, F. J., & Matthews, D. H. (1963). Magnetic Anomalies Over Oceanic Ridges. Nature, 199(4897), 947–949. doi:10.1038/199947a0

Voosen, P. (2022). Bogs, lakebeds, and sea floors compete to become Anthropocene's 'golden spike' [Data set].
doi:10.1126/science.abq8466

Walker, M. P. (2017). Why we sleep: Unlocking the power of sleep and dreams (First Scribner hardcover edition). Scribner, an imprint of Simon & Schuster, Inc.

Webb, B. M., & Head, J. W., III. (2002). Noachian Tectonics of Syria Planum and the Thaumasia Plateau. 1358.

Weidenschilling, S. J. (1977). The distribution of mass in the planetary system and solar nebula. Astrophysics and Space Science, 51(1), 153–158.
doi:10.1007/BF00642464

West, G. B. (2017). Scale: The universal laws of growth, innovation, sustainability, and the pace of life in organisms, cities, economies, and companies. Penguin Press.

Yin, A. (2012). An episodic slab-rollback model for the origin of the Tharsis rise on Mars: Implications for initiation of local plate subduction and final unification of a kinematically linked global plate-tectonic network on Earth. Lithosphere, 4(6), 553–593. doi:10.1130/L195.1

I'm proud of you.

I'm glad you're here.

and I love you.

.

Made in the USA
Las Vegas, NV
03 February 2024

85243619R00118